Seventh International Workshop on Health Text Mining and Information Analysis (LOUHI 2016)

Held at the Conference on Empirical Methods in Natural Language Processing (EMNLP 2016)

Austin, Texas, USA
5 November 2016

ISBN: 978-1-5108-3152-0

Preface

The Seventh International Workshop on Health Text Mining and Information Analysis provides an interdisciplinary forum for researchers interested in automated processing of health documents. Health documents encompass electronic health records, clinical guidelines, spontaneous reports for pharmacovigilance, biomedical literature, health forums/blogs or any other type of health-related documents. The LOUHI workshop series fosters interactions between the Computational Linguistics, Medical Informatics and Artificial Intelligence communities. Following the six previous edition of the workshop which were co-located with SMBM 2008 in Turku, Finland, with NAACL 2010 in Los Angeles, California, with Artificial Intelligence in Medicine (AIME 2011) in Bled, Slovenia, during NICTA Techfest 2013 in Sydney, Australia, co-located with EACL 2014 in Gothenburg, Sweden, and with EMNLP 2015 in Lisbon, Portugal, this workshop is co-located this year with EMNLP 2016 in Austin, Texas.

The aim of the LOUHI 2016 workshop is to bring together research work on topics related to health documents, particularly emphasizing multidisciplinary aspects of health documentation and the interplay between nursing and medical sciences, information systems, computational linguistics and computer science. The topics include, but are not limited to, the following Natural Language Processing techniques and related areas:

- Techniques supporting information extraction, e.g. named entity recognition, negation and uncertainty detection

- Classification and text mining applications (e.g. diagnostic classifications such as ICD-10 and nursing intensity scores) and problems (e.g. handling of unbalanced data sets)

- Text representation, including dealing with data sparsity and dimensionality issues

- Domain adaptation, e.g. adaptation of standard NLP tools (incl. tokenizers, PoS-taggers, etc) to the medical domain

- Information fusion, i.e. integrating data from various sources, e.g. structured and narrative documentation

- Unsupervised methods, including distributional semantics

- Evaluation, gold/reference standard construction and annotation

- Syntactic, semantic and pragmatic analysis of health documents

- Anonymization/de-identification of health records and ethics

- Supporting the development of medical terminologies and ontologies

- Individualization of content, consumer health vocabularies, summarization and simplification of text

- NLP for supporting documentation and decision making practices

- Predictive modeling of adverse events, e.g. adverse drug events and hospital acquired infections

The call for papers encouraged authors to submit papers describing substantial and completed work but also focus on a contribution, a negative result, a software package or work in progress. We also encouraged to report work on low-resourced languages, addressing the challenges of data sparsity and language characteristic diversity.

We received 21 submissions. Each submission went through a double-blind review process which involved three program committee members. Based on comments and rankings supplied by the reviewers, we accepted 13 papers. The overall acceptance rate is 62%. During the workshop, 6 papers will be presented orally, and 7 papers will be presented as posters.

Our special thanks go to Nigel Collier for accepting to give an invited talk.

Finally, we would like to thank the members of the program committee for the quality of theirs reviews in a very short period, and the authors for their submissions and the quality of their work.

Cyril Grouin, Thierry Hamon, Aurélie Névéol, Pierre Zweigenbaum.

Organizers:

Cyril Grouin, CNRS, LIMSI (France)
Thierry Hamon, Université Paris 13, CNRS, LIMSI (France)
Aurélie Névéol, CNRS, LIMSI (France)
Pierre Zweigenbaum, CNRS, LIMSI (France)

Program Committee:

Sophia Ananiadou, University of Manchester (UK)
Sabine Bergler, Concordia University (Canada)
Thomas Brox Røst, Norwegian University of Science and Technology (Norway)
Kevin B Cohen, University of Colorado/School of Medicine (USA)
Hercules Dalianis, Stockholm University (Sweden)
Louise Deléger, INRA (France)
Filip Ginter, University of Turku (Finland)
Natalia Grabar, CNRS UMR 8163, STL Université de Lille 3 (France)
Gintaré Grigonyté, Stockholm University (Sweden)
Aron Henriksson, Stockholm University (Sweden)
Antonio Jimeno Yepes, IBM Research (Australia)
Jussi Karlgren, KTH, Royal Institute of Technology (Sweden)
Dimitrios Kokkinakis, University of Gothenburg (Sweden)
Maria Kvist, Stockholm University (Sweden)
Alberto Lavelli, Fondazione Bruno Kessler (Italy)
David Martinez, University of Melbourne and MedWhat.com (Australia)
Beáta Megyesi, Uppsala University (Sweden)
Marie-Jean Meurs, UQAM & Concordia University (Canada)
Fleur Mougin, Université de Bordeaux, ERIAS, Centre INSERM U897, ISPED (France)
Danielle L Mowery, University of Utah (USA)
Henning Müller, University of Applied Sciences Western Switzerland (Switzerland)
Mariana Lara Neves, Hasso-Plattner-Institute at the University of Potsdam (Germany)
Jong C. Park, KAIST Computer Science (Korea)
Rezarta Islamaj-Dogan, NIH/NLM/NCBI (USA)
Tapio Salakoski, University of Turku (Finland)
Stefan Schulz, Graz General Hospital and University Clinics (Austria)
Isabel Segura-Bedmar, Universidad Carlos III de Madrid (Spain)
Maria Skeppstedt, Stockholm University (Sweden)
Hanna Suominen, NICTA (Australia)
Suzanne Tamang, Stanford University School of Medicine (USA)
Özlem Uzuner, MIT (USA)
Sumithra Velupillai, Stockholm University (Sweden)
Karin Verspoor, University of Melbourne (Australia)
Mats Wirén, Stockholm University, Stockholm (Sweden)

Invited Speaker:

Nigel Collier, University of Cambridge, UK

Table of Contents

An Investigation of Recurrent Neural Architectures for Drug Name Recognition
Raghavendra Chalapathy, Ehsan Zare Borzeshi and Massimo Piccardi 1

Clinical Text Prediction with Numerically Grounded Conditional Language Models
Georgios Spithourakis, Steffen Petersen and Sebastian Riedel 6

Modelling Radiological Language with Bidirectional Long Short-Term Memory Networks
Savelie Cornegruta, Robert Bakewell, Samuel Withey and Giovanni Montana 17

Data Resource Acquisition from People at Various Stages of Cognitive Decline – Design and Exploration Considerations
Dimitrios Kokkinakis, Kristina Lundholm Fors and Arto Nordlund 28

Analysis of Anxious Word Usage on Online Health Forums
Nicolas Rey-Villamizar, Prasha Shrestha, Farig Sadeque, Steven Bethard, Ted Pedersen, Arjun Mukherjee and Thamar Solorio 37

Retrofitting Word Vectors of MeSH Terms to Improve Semantic Similarity Measures
Zhiguo Yu, Trevor Cohen, Byron Wallace, Elmer Bernstam and Todd Johnson 43

Unsupervised Resolution of Acronyms and Abbreviations in Nursing Notes Using Document-Level Context Models
Katrin Kirchhoff and Anne M. Turner 52

Low-resource OCR error detection and correction in French Clinical Texts
Eva D'hondt, Cyril Grouin and Brigitte Grau 61

Citation Analysis with Neural Attention Models
Tsendsuren Munkhdalai, John Lalor and Hong Yu 69

Replicability of Research in Biomedical Natural Language Processing: a pilot evaluation for a coding task
Aurelie Neveol, Kevin Cohen, Cyril Grouin and Aude Robert 78

NLP and Online Health Reports: What do we say and what do we mean?
Nigel Collier 85

Leveraging coreference to identify arms in medical abstracts: An experimental study
Elisa Ferracane, Iain Marshall, Byron C. Wallace and Katrin Erk 86

Hybrid methods for ICD-10 coding of death certificates
Pierre Zweigenbaum and Thomas Lavergne 96

Exploring Query Expansion for Entity Searches in PubMed
Chung-Chi Huang and Zhiyong Lu 106

Workshop Program

Saturday, November 5, 2016

09:00–10:15   **Session I - Machine-Learning**

09:00–09:25   *An Investigation of Recurrent Neural Architectures for Drug Name Recognition*
Raghavendra Chalapathy, Ehsan Zare Borzeshi and Massimo Piccardi

09:25–09:50   *Clinical Text Prediction with Numerically Grounded Conditional Language Models*
Georgios Spithourakis, Steffen Petersen and Sebastian Riedel

09:50–10:15   *Modelling Radiological Language with Bidirectional Long Short-Term Memory Networks*
Savelie Cornegruta, Robert Bakewell, Samuel Withey and Giovanni Montana

10:15–10:30   **Session II - Boosters**

10:30–11:00   *Coffee Break*

11:00–12:30   **Session III - Posters**

Data Resource Acquisition from People at Various Stages of Cognitive Decline – Design and Exploration Considerations
Dimitrios Kokkinakis, Kristina Lundholm Fors and Arto Nordlund

Analysis of Anxious Word Usage on Online Health Forums
Nicolas Rey-Villamizar, Prasha Shrestha, Farig Sadeque, Steven Bethard, Ted Pedersen, Arjun Mukherjee and Thamar Solorio

Retrofitting Word Vectors of MeSH Terms to Improve Semantic Similarity Measures
Zhiguo Yu, Trevor Cohen, Byron Wallace, Elmer Bernstam and Todd Johnson

Unsupervised Resolution of Acronyms and Abbreviations in Nursing Notes Using Document-Level Context Models
Katrin Kirchhoff and Anne M. Turner

Low-resource OCR error detection and correction in French Clinical Texts
Eva D'hondt, Cyril Grouin and Brigitte Grau

Citation Analysis with Neural Attention Models
Tsendsuren Munkhdalai, John Lalor and Hong Yu

Replicability of Research in Biomedical Natural Language Processing: a pilot evaluation for a coding task
Aurelie Neveol, Kevin Cohen, Cyril Grouin and Aude Robert

12:30–14:00   ***Lunch break***

14:00–15:30   **Session IV - Invited talk**

14:00–15:30   *NLP and Online Health Reports: What do we say and what do we mean?*
Nigel Collier

15:30–16:00   ***Coffee Break***

16:00–17:15   **Session V - NLP for literature and clinical documents**

16:00–16:25   *Leveraging coreference to identify arms in medical abstracts: An experimental study*
Elisa Ferracane, Iain Marshall, Byron C. Wallace and Katrin Erk

16:25–16:50   *Hybrid methods for ICD-10 coding of death certificates*
Pierre Zweigenbaum and Thomas Lavergne

16:50–17:15   *Exploring Query Expansion for Entity Searches in PubMed*
Chung-Chi Huang and Zhiyong Lu

An Investigation of Recurrent Neural Architectures
for Drug Name Recognition

Raghavendra Chalapathy
University of Sydney
J12/1 Cleveland St
Darlington NSW 2008
rcha9612@uni.sydney.edu.au

Ehsan Zare Borzeshi
Capital Markets CRC
3/55 Harrington St
Sydney NSW 2000
ezborzeshi@cmcrc.com

Massimo Piccardi
University of Technology Sydney
PO Box 123
Broadway NSW 2007
Massimo.Piccardi@uts.edu.au

Abstract

Drug name recognition (DNR) is an essential step in the Pharmacovigilance (PV) pipeline. DNR aims to find drug name mentions in unstructured biomedical texts and classify them into predefined categories. State-of-the-art DNR approaches heavily rely on hand-crafted features and domain-specific resources which are difficult to collect and tune. For this reason, this paper investigates the effectiveness of contemporary recurrent neural architectures - the Elman and Jordan networks and the bidirectional LSTM with CRF decoding - at performing DNR straight from the text. The experimental results achieved on the authoritative SemEval-2013 Task 9.1 benchmarks show that the bidirectional LSTM-CRF ranks closely to highly-dedicated, hand-crafted systems.

1 Introduction

Pharmacovigilance (PV) is defined by the World Health Organization as the science and activities concerned with the detection, assessment, understanding and prevention of adverse effects of drugs or any other drug-related problems. Drug name recognition (DNR) is a fundamental step in the PV pipeline, similarly to the well-studied Named Entity Recognition (NER) task for general natural language processing (NLP). DNR aims to find drug mentions in unstructured biomedical texts and classify them into predefined categories in order to link drug names with their effects and explore drug-drug interactions (DDIs). Conventional approaches to DNR sub-divide as rule-based, dictionary-based and machine learning-based. Intrinsically, rule-based systems are hard to scale, time-consuming to assemble and ineffective in the presence of informal sentences and abbreviated phrases. Dictionary-based systems identify drug names by matching text chunks against drug dictionaries. These systems typically achieve high precision, but suffer from low recall (i.e., they miss a significant number of mentions) due to spelling errors or drug name variants not present in the dictionaries (Liu et al., 2015a). Conversely, machine-learning approaches have the potential to overcome all these limitations since their foundations are intrinsically robust to variants. The current state-of-the-art machine learning approaches follow a two-step process of feature engineering and classification (Segura-Bedmar et al., 2015; Abacha et al., 2015; Rocktäschel et al., 2013). Feature engineering refers to the task of representing text by dedicated numeric vectors using domain knowledge. Similarly to the design of rule-based systems, this task requires much expert knowledge, is typically challenging and time-consuming, and has a major impact on the final accuracy. For this reason, this paper explores the performance of contemporary recurrent neural networks (RNNs) at providing end-to-end DNR straight from text, without any manual feature engineering stage. The tested RNNs include the popular Elman and Jordan networks and the bidirectional long short-term memory (LSTM) with decoding provided by a conditional random field (CRF) (Elman, 1990; Jordan, 1986; Lample et al., 2016; Collobert et al., 2011). The experimental results over the SemEval-2013 Task 9.1 benchmarks show an interesting accuracy from the LSTM-CRF

1

Proceedings of the Seventh International Workshop on Health Text Mining and Information Analysis (LOUHI), pages 1–5,
Austin, TX, November 5, 2016. ©2016 Association for Computational Linguistics

that exceeds that of various manually-engineered systems and approximates the best result in the literature.

2   Related Work

Most of the research on drug name recognition to date has focussed on domain-dependent aspects and specialized text features. The benefit of leveraging such tailored features was made evident by the results from the SemEval-2013 Task 9.1 (Recognition and classification of pharmacological substances, known as DNR task) challenge. The system that ranked first, WBI-NER (Rocktäschel et al., 2013), adopted very specialized features derived from an improved version of the ChemSpot tool (Rocktäschel et al., 2012), a collection of drug dictionaries and ontologies. Similarly, many other recent approaches (Abacha et al., 2015; Liu et al., 2015b; Segura-Bedmar et al., 2015) have been based on various combinations of general and domain-specific features. In the broader field of machine learning, the recent years have witnessed a rapid proliferation of deep neural networks, with unprecedented results in tasks as diverse as visual, speech and named-entity recognition (Hinton et al., 2012; Krizhevsky et al., 2012; Lample et al., 2016). One of the main advantages of neural networks is that they can learn the feature representations automatically from the data, thus avoiding the laborious feature engineering stage (Mesnil et al., 2015; Lample et al., 2016). Given these promising results, the main goal of this paper is to provide the first performance investigation of popular RNNs such as the Elman and Jordan networks and the bidirectional LSTM-CRF over DNR tasks.

3   The Proposed Approach

DNR can be formulated as a joint segmentation and classification task over a predefined set of classes. As an example, consider the input sentence provided in Table 1. The notation follows the widely adopted in/out/begin (IOB) entity representation with, in this instance, *Cimetidine* as the drug, *ALFENTA* as the brand, and words *volatile inhalation anesthetics* together as the group. In this paper, we approach the DNR task by recurrent neural networks and we therefore provide a brief description hereafter. In an RNN, each word in the input sentence is first mapped to a random real-valued vector of arbitrary dimension, d. Then, a measurement for the word, noted as $x(t)$, is formed by concatenating the word's own vector with a window of preceding and following vectors (the "context"). An example of input vector with a context window of size $s = 3$ is:

$$w_3(t) = [Cimetidine, \textbf{reduces}, effect],$$
$$`reduces' \rightarrow x_{reduces} \in \mathbb{R}^d,$$
$$`Cimetidine' \rightarrow x_{Cimetidine} \in \mathbb{R}^d, \quad (1)$$
$$`effect' \rightarrow x_{effect} \in \mathbb{R}^d,$$
$$x(t) = [x_{Cimetidine}, x_{\textbf{reduces}}, x_{effect}] \in \mathbb{R}^{3d}$$

where $w_3(t)$ is the context window centered around the t-th word, $'reduces'$, and x_{word} represents the numerical vector for *word*.

For the Elman network, both $x(t)$ and the output from the hidden layer at time $t - 1$, $h(t - 1)$, are input into the hidden layer for frame t. The recurrent connection from the past time frame enables a short-term memory, while hidden-to-hidden neuron connections make the network Turing-complete. This architecture, common in RNNs, is suitable for prediction of sequences. Formally, the hidden layer is described as:

$$h(t) = f(U \bullet x(t) + V \bullet h(t - 1)) \qquad (2)$$

where U and V are randomly-initialized weight matrices between the input and the hidden layer, and between the past and current hidden layers, respectively. Function $f(\cdot)$ is the sigmoid function:

$$f(x) = \frac{1}{1 + e^{-x}} \qquad (3)$$

that adds non-linearity to the layer. Eventually, $h(t)$ is input in the output layer:

$$y(t) = g(W \bullet h(t)), \text{ with } g(z_m) = \frac{e^{z_m}}{\Sigma_{k=1}^{K} e^{z_k}} \qquad (4)$$

and convolved with the output weight matrix, W. The output is normalized by a multi-class logistic function, $g(\cdot)$, to become a proper probability over the class set. The output dimensionality is therefore determined by the number of entity classes (i.e., 4 for the DNR task).

The Jordan network is very similar to the Elman network, except that the feedback is sourced

Sentence	*Cimetidine*	*reduces*	*clearance*	*of*	*ALFENTA*	*and*	*volatile*	*inhalation*	*anesthetics*
Entity class	*B-drug*	*O*	*O*	*O*	*B-brand*	*O*	*B-group*	*I-group*	*I-group*

Table 1: Example sentence in a DNR task with entity classes represented in IOB format.

	DDI-DrugBank		DDI-MedLine	
	Training+Test for DDI task	Test for DNR	Training+Test for DDI task	Test for DNR
documents	730	54	175	58
sentences	6577	145	1627	520
drug_n	124	6	520	115
group	3832	65	234	90
brand	1770	53	36	6
drug	9715	180	1574	171

Table 2: Statistics of training and test datasets used for SemEval-2013 Task 9.1.

from the output layer rather than the previous hidden layer:

$$h(t) = f(U \bullet x(t) + V \bullet y(t-1)). \qquad (5)$$

Although the Elman and Jordan networks can learn long-term dependencies, their exponential decay biases them toward their most recent inputs (Bengio et al., 1994). The LSTM was designed to overcome this limitation by incorporating a gated memory-cell to capture long-range dependencies within the data (Hochreiter and Schmidhuber, 1997). In the bidirectional LSTM, for any given sentence, the network computes both a left, $\overrightarrow{h}(t)$, and a right, $\overleftarrow{h}(t)$, representations of the sentence context at every input, $x(t)$. The final representation is created by concatenating them as $h(t) = [\overrightarrow{h}(t); \overleftarrow{h}(t)]$. All these networks utilize the $h(t)$ layer as an implicit feature for entity class prediction: although this model has proved effective in many cases, it is not able to provide joint decoding of the outputs in a Viterbi-style manner (e.g., an I-group cannot follow a B-brand; etc). Thus, another modification to the bidirectional LSTM is the addition of a conditional random field (CRF) (Lafferty et al., 2001) as the output layer to provide optimal sequential decoding. The resulting network is commonly referred to as the bidirectional LSTM-CRF (Lample et al., 2016).

4 Experiments

4.1 Datasets

The DDIExtraction 2013 shared task challenge from SemEval-2013 Task 9.1 (Segura-Bedmar et al., 2013) has provided a benchmark corpus for DNR and DDI extraction. The corpus contains manually-annotated pharmacological substances and drug-drug interactions (DDIs) for a total of $18,502$ pharmacological substances and $5,028$ DDIs. It collates two distinct datasets: DDI-DrugBank and DDI-MedLine (Herrero-Zazo et al., 2013). Table 2 summarizes the basic statistics of the training and test datasets used in our experiments. For proper comparison, we follow the same settings as (Segura-Bedmar et al., 2015), using the training data of the DNR task along with the test data for the DDI task for training and validation of DNR. We split this joint dataset into a training and validation sets with approximately 70% of sentences for training and the remaining for validation.

4.2 Evaluation Methodology

Our models have been blindly evaluated on unseen DNR test data using the *strict* evaluation metrics. With this evaluation, the predicted entities have to match the ground-truth entities exactly, both in boundary and class. To facilitate the replication of our experimental results, we have used a publicly-available library for the implementation[1] (i.e., the Theano neural network toolkit (Bergstra et al., 2010)). The experiments have been run over a range of values for the hyper-parameters, using the validation set for selection (Bergstra and Bengio, 2012). The hyper-parameters include the number of hidden-layer nodes, $H \in \{25, 50, 100\}$, the context window size, $s \in \{1, 3, 5\}$, and the embedding dimension, $d \in \{50, 100, 300, 500, 1000\}$. Two addi-

---

[1] https://github.com/raghavchalapathy/dnr

Methods	DDI-DrugBank			DDI-MedLine		
	Precision	Recall	F_1 Score	Precision	Recall	F_1 Score
WBI-NER (Rocktäschel et al., 2013)	88.00	87.00	87.80	61.00	56.00	58.10
Hybrid-DDI (Abacha et al., 2015)	93.00	70.00	80.00	74.00	25.00	37.00
Word2Vec+DINTO (Segura-Bedmar et al., 2015)	69.00	82.00	75.00	65.00	51.00	57.00
Elman RNN	79.91	60.91	69.13	43.23	33.56	37.78
Jordan RNN	77.59	60.91	68.25	59.47	30.20	40.06
Bidirectional LSTM-CRF	87.07	83.39	85.19	52.93	52.57	52.75

Table 3: Performance comparison between the recurrent neural networks (bottom three lines) and state-of-the-art systems (top three lines) over the SemEval-2013 Task 9.1.

	Entities	DDI-DrugBank			DDI-MedLine		
		Precision	Recall	F_1 Score	Precision	Recall	F_1 Score
Bidirectional LSTM-CRF	group	76.92	90.91	83.33	59.52	53.76	56.50
	drug	90.59	84.62	87.50	65.22	61.05	63.06
	brand	91.30	79.25	84.85	0.0	0.0	0.0
	drug_n	0.0	0.0	0.0	40.20	45.45	42.67

Table 4: SemEval-2013 Task 9.1 results by entity for the bidirectional LSTM-CRF.

tional parameters, the learning and drop-out rates, were sampled from a uniform distribution in the range $[0.05, 0.1]$. The embedding and initial weight matrices were all sampled from the uniform distribution within range $[-1, 1]$. Early training stopping was set to 100 epochs to mollify over-fitting, and the model that gave the best performance on the validation set was retained. The accuracy is reported in terms of micro-average F_1 score computed using the CoNLL score function (Nadeau and Sekine, 2007).

4.3 Results and Analysis

Table 3 shows the performance comparison between the explored RNNs and state-of-the-art DNR systems. As an overall note, the RNNs have not reached the same accuracy as the top system, WBI-NER (Rocktäschel et al., 2013). However, the bidirectional LSTM-CRF has achieved the second-best score on DDI-DrugBank and the third-best on DDI-MedLine. These results seem interesting on the ground that the RNNs provide DNR straight from text rather than from manually-engineered features. Given that the RNNs learn entirely from the data, the better performance over the DDI-DrugBank dataset is very likely due to its larger size. Accordingly, it is reasonable to expect higher relative performance should larger corpora become available in the future. Table 4 also breaks down the results by entity class for the bidirectional LSTM-CRF. The low score on the *brand* class for DDI-MedLine and on the *drug_n* class (i.e., active substances not approved for human use) for DDI-DrugBank are likely attributable to the very small sample size (Table 2). This issue is also shared by the state-of-the-art DNR systems.

5 Conclusion

This paper has investigated the effectiveness of recurrent neural architectures, namely the Elman and Jordan networks and the bidirectional LSTM-CRF, for drug name recognition. The most appealing feature of these architectures is their ability to provide end-to-end recognition straight from text, sparing effort from laborious feature construction. To the best of our knowledge, ours is the first paper to explore RNNs for entity recognition from pharmacological text. The experimental results over the SemEval-2013 Task 9.1 benchmarks look promising, with the bidirectional LSTM-CRF ranking closely to the state of the art. A potential way to further improve its performance would be to initialize its training with unsupervised word embeddings such as Word2Vec (Mikolov et al., 2013) and GloVe (Pennington et al., 2014). This approach has proved effective in many other domains and still dispenses with expert annotation effort; we plan this exploration for the near future.

References

Asma Ben Abacha, Md Faisal Mahbub Chowdhury, Aikaterini Karanasiou, Yassine Mrabet, Alberto Lavelli, and Pierre Zweigenbaum. 2015. Text mining for pharmacovigilance: Using machine learning for drug name recognition and drug–drug interaction extraction and classification. *Journal of Biomedical Informatics*, 58:122–132.

Yoshua Bengio, Patrice Simard, and Paolo Frasconi. 1994. Learning long-term dependencies with gradient descent is difficult. *IEEE Transactions on Neural Networks*, 5(2):157–166.

James Bergstra and Yoshua Bengio. 2012. Random search for hyper-parameter optimization. *Journal of Machine Learning Research*, 13:281–305.

James Bergstra, Olivier Breuleux, Frédéric Bastien, Pascal Lamblin, Razvan Pascanu, Guillaume Desjardins, Joseph Turian, David Warde-Farley, and Yoshua Bengio. 2010. Theano: A CPU and GPU math compiler in Python. In *The 9th Python in Science Conference*, pages 1–7.

Ronan Collobert, Jason Weston, Léon Bottou, Michael Karlen, Koray Kavukcuoglu, and Pavel Kuksa. 2011. Natural language processing (almost) from scratch. *Journal of Machine Learning Research*, 12:2493–2537.

Jeffrey L Elman. 1990. Finding structure in time. *Cognitive Science*, 14(2):179–211.

María Herrero-Zazo, Isabel Segura-Bedmar, Paloma Martínez, and Thierry Declerck. 2013. The DDI corpus: An annotated corpus with pharmacological substances and drug–drug interactions. *Journal of Biomedical Informatics*, 46(5):914–920.

Geoffrey Hinton, Li Deng, Dong Yu, George E Dahl, Abdel-rahman Mohamed, Navdeep Jaitly, Andrew Senior, Vincent Vanhoucke, Patrick Nguyen, Tara N Sainath, et al. 2012. Deep neural networks for acoustic modeling in speech recognition: The shared views of four research groups. *IEEE Signal Processing Magazine*, 29(6):82–97.

Sepp Hochreiter and Jürgen Schmidhuber. 1997. Long short-term memory. *Neural Computation*, 9(8):1735–1780.

Michael I. Jordan. 1986. Serial order: A parallel distributed processing approach. Technical report, San Diego: University of California, Institute for Cognitive Science.

Alex Krizhevsky, Ilya Sutskever, and Geoffrey E Hinton. 2012. Imagenet classification with deep convolutional neural networks. In *NIPS*, pages 1097–1105.

John Lafferty, Andrew McCallum, and Fernando Pereira. 2001. Conditional random fields: Probabilistic models for segmenting and labeling sequence data. In *ICML*, pages 282–289.

Guillaume Lample, Miguel Ballesteros, Sandeep Subramanian, Kazuya Kawakami, and Chris Dyer. 2016. Neural architectures for named entity recognition. In *NAACL-HLT*.

Shengyu Liu, Buzhou Tang, Qingcai Chen, and Xiaolong Wang. 2015a. Drug name recognition: Approaches and resources. *Information*, 6(4):790–810.

Shengyu Liu, Buzhou Tang, Qingcai Chen, Xiaolong Wang, and Xiaoming Fan. 2015b. Feature engineering for drug name recognition in biomedical texts: Feature conjunction and feature selection. *Computational and Mathematical Methods in Medicine*, 2015:1–9.

Grégoire Mesnil, Yann Dauphin, Kaisheng Yao, Yoshua Bengio, Li Deng, Dilek Hakkani-Tur, Xiaodong He, Larry Heck, Gokhan Tur, Dong Yu, et al. 2015. Using recurrent neural networks for slot filling in spoken language understanding. *IEEE/ACM Transactions on Audio, Speech, and Language Processing*, 23(3):530–539.

Tomas Mikolov, Ilya Sutskever, Kai Chen, Greg S Corrado, and Jeff Dean. 2013. Distributed representations of words and phrases and their compositionality. In *NIPS*, pages 3111–3119.

David Nadeau and Satoshi Sekine. 2007. A survey of named entity recognition and classification. *Linguisticae Investigationes*, 30(1):3–26.

Jeffrey Pennington, Richard Socher, and Christopher D. Manning. 2014. GloVe: Global vectors for word representation. In *EMNLP*, pages 1532–1543.

Tim Rocktäschel, Michael Weidlich, and Ulf Leser. 2012. ChemSpot: A hybrid system for chemical named entity recognition. *Bioinformatics*, 28(12):1633–1640.

Tim Rocktäschel, Torsten Huber, Michael Weidlich, and Ulf Leser. 2013. WBI-NER: The impact of domain-specific features on the performance of identifying and classifying mentions of drugs. In *The 7th International Workshop on Semantic Evaluation*, pages 356–363.

Isabel Segura-Bedmar, Paloma Martínez, and María Herrero Zazo. 2013. Semeval-2013 task 9: Extraction of drug-drug interactions from biomedical texts (DDIExtraction 2013). In *The 7th International Workshop on Semantic Evaluation*.

Isabel Segura-Bedmar, Víctor Suárez-Paniagua, and Paloma Martínez. 2015. Exploring word embedding for drug name recognition. In *The 6th International Workshop on Health Text Mining and Information Analysis*, page 64.

Clinical Text Prediction
with Numerically Grounded Conditional Language Models

Georgios P. Spithourakis
Department of Computer Science
University College London
g.spithourakis@cs.ucl.ac.uk

Steffen E. Petersen
William Harvey Research Institute
Queen Mary University of London
s.e.petersen@qmul.ac.uk

Sebastian Riedel
Department of Computer Science
University College London
s.riedel@cs.ucl.ac.uk

Abstract

Assisted text input techniques can save time and effort and improve text quality. In this paper, we investigate how grounded and conditional extensions to standard neural language models can bring improvements in the tasks of word prediction and completion. These extensions incorporate a structured knowledge base and numerical values from the text into the context used to predict the next word. Our automated evaluation on a clinical dataset shows extended models significantly outperform standard models. Our best system uses both conditioning and grounding, because of their orthogonal benefits. For word prediction with a list of 5 suggestions, it improves recall from 25.03% to 71.28% and for word completion it improves keystroke savings from 34.35% to 44.81%, where theoretical bound for this dataset is 58.78%. We also perform a qualitative investigation of how models with lower perplexity occasionally fare better at the tasks. We found that at test time numbers have more influence on the document level than on individual word probabilities.

1 Introduction

Text prediction is the task of suggesting the next word, phrase or sentence while the user is typing. It is an assisted data entry function that aims to save time and effort by reducing the number of keystrokes needed and to improve text quality by preventing misspellings, promoting adoption of standard terminologies and allowing for exploration of the vocabulary (Sevenster and Aleksovski, 2010; Sevenster et al., 2012).

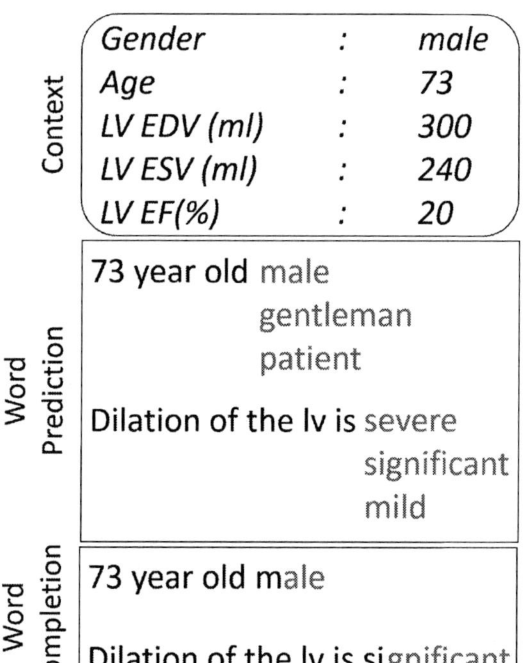

Figure 1: Word prediction and completion tasks. A system makes suggestions (in grey) for the next word and to complete a word as it is being typed, respectively. The context is often relevant to the quality of the suggestions.

Text prediction originated in augmentative and alternative communication (AAC) to increase text generation rates for people with motor or speech impairments (Beukelman and Mirenda, 2005). Its scope has been extended to a gamut of applications, such as data entry in mobile devices (Dunlop and Crossan, 2000), interactive machine translation (Foster et al., 2002), search term auto-

Proceedings of the Seventh International Workshop on Health Text Mining and Information Analysis (LOUHI), pages 6–16,
Austin, TX, November 5, 2016. ©2016 Association for Computational Linguistics

completion (Bast and Weber, 2006) and assisted clinical report compilation (Eng and Eisner, 2004; Cannataro et al., 2012).

In this paper, we explore the tasks of word prediction, where a system displays a list of suggestions for the next word before the user starts typing it, and word completion, where the system suggests a single possible completion for the word, while the user is typing its characters. The former task is relevant when the user has not yet made a firm decision about the intended word, thus any suggestions can have a great impact in the content of the final document. In the latter case, the user is thinking of a particular word that they want to input and the system's goal is to help them complete the word as quickly as possible. Figure 1 shows examples for both tasks.

Often, the user's goal is to compose a document describing a particular situation, e.g. a clinical report about a patient's condition. An intelligent predictive system should be able to account for such contextual information in order to improve the quality of its suggestions. Challenges to modelling structured contexts include mixed types of values for the different fields an schema inconsistencies across the entries of the structure. We address these issues by employing numerically grounded conditional language models (Spithourakis et al., 2016).

The contribution of this work is twofold. First, we show that conditional and numerically grounded models can achieve significant improvements over standard language models in the tasks of word prediction and completion. Our best model with a list of 5 suggestions raises recall from 25.03% to 71.28% and keystroke savings from 34.35% to 44.81%. Second, we investigate in depth the behaviour of such models and their sensitivity to the numerical values in the text. We find that the grounded probability for the whole document is more sensitive to numerical configurations than the probabilities of individual words.

2   Related Work

There have been several applications of text prediction systems in the clinical domain. Word completion has been a feature of discharge summary (Chen et al., 2012), brain MRI report (Cannataro et al., 2012) and radiology report (Eng and Eisner, 2004)

compilation systems. Aiming towards clinical document standardisation, Sirel (2012) adopted the ICD-10 medical classification codes as a lexical resource and Lin et al. (2014) built a semi-automatic annotation tool to generate entry-level interoperable clinical documents.

Hua et al. (2014) reported 13.0% time reduction and 3.9% increase of response accuracy in a data entry task. Gong et al. (2016) found a performance of 87.1% for keystroke savings, a 70.5% increase in text generation rate, a 34.1% increase in reporting comprehensiveness and a 14.5% reduction in non-adherence to fields when reporting on patient safety event. In non-clinical applications, a survey of text prediction systems (Garay-Vitoria and Abascal, 2006) reports keystroke savings ranging from 29% to 56%.

The context provided to the predictive system can have a significant effect on its performance. Fazly and Hirst (2003) and Van Den Bosch and Bogers (2008) obtained significantly better results for word completion by considering not only the prefix of the current word but also previous words and characters, respectively. Wandmacher and Antoine (2008) explored methods to integrate n-gram language models with semantic information and Trnka (2008) used topic-adapted language models for word prediction. More recently, Ghosh et al. (2016) incorporated sentence topics as contextual features into a neural language model and reported perplexity improvements in a word prediction task. None of these systems considers structured background information or numerical values from the text as additional context.

The motivation to include this information as context to text prediction system is based on the importance of numerical quantities to textual entailment systems (Roy et al., 2015; Sammons et al., 2010; MacCartney and Manning, 2008; De Marneffe et al., 2008). In medical communications, sole use of verbal specifications (e.g. adjectives and adverbs) has been associated with less precise understanding of frequencies (Nakao and Axelrod, 1983) and probabilities (Timmermans, 1994). A combination of structured data and free text is deemed more suitable for communicating clinical information (Lovis et al., 2000).

Language models have been an integral part of

text prediction systems (Bickel et al., 2005; Wandmacher and Antoine, 2008; Trnka, 2008; Ghosh et al., 2016). Several tasks call for generative language models that have been conditioned on various contexts, e.g. foreign language text for machine translation (Cho et al., 2014), images (Vinyals et al., 2015; Donahue et al., 2015) and videos (Yao et al., 2015) for captioning, etc. Grounded language models represent the relationship between words and the non-linguistic context they refer to. Grounding can help learn better representations for the atoms of language and their interactions. Previous work grounds language on vision (Bruni et al., 2014; Silberer and Lapata, 2014), audio (Kiela and Clark, 2015), video (Fleischman and Roy, 2008) and the olfactory perception (Kiela et al., 2015). Spithourakis et al. (2016) use numerically grounded language models and language models conditioned on a lexicalised knowledge base for the tasks of semantic error detection and correction. We directly use their models to perform word prediction and completion.

3 Methodology

In this section we present a solution to the word prediction and completion tasks (Subsection 3.1). Then, we discuss how language models, which can be grounded on numeric quantities mentioned in the text and/or conditioned on external context can be used in our framework (Subsection 3.2). Finally, we describe our automated evaluation process and various evaluation metrics for the two tasks (Subsection 3.3).

3.1 Word prediction and completion

Let $\{w_1, ..., w_T\}$ denote a document, where w_t is the word at position t. Documents are often associated with external context that can be structured (e.g. a knowledge base) or unstructured (e.g. other documents). Let's consider the case where our context is a knowledge base (KB), that is a set of tuples of the form $< attribute, value >$, where attributes are defined by the KB schema. Different attributes might take values from different domains, e.g. strings, binary values, real numbers etc., and some of the values might be missing.

In the word prediction task, the system presents a ranked list of suggestions for the next word to

---

Algorithm 1 Word completion

Input: $\mathcal{V}$ is set of vocabulary words, $scorer$ returns score for word in current position
Output: next word to be written
1: **function** COMPLETEWORD($\mathcal{V}, scorer$)
2:   $prefix \leftarrow$ ''
3:   $lexicon \leftarrow \mathcal{V}$
4:   **loop**
5:    $lexicon \leftarrow \{$tokens in $lexicon$ starting with $prefix\}$
6:    $best \leftarrow \underset{token \in lexicon}{\mathrm{argmax}}\ scorer(token)$
7:    Display $best$
8:    $char \leftarrow$ read next char
9:    **if** $char = TAB$ **then**
10:     **return** best   ▷ Auto-complete
11:    **else if** $char = WHITESPACE$ **then**
12:     **return** $prefix$ ▷ Next word
13:    **else**
14:     $prefix \leftarrow prefix + char$ ▷ Append
15:    **end if**
16:   **end loop**
17: **end function**

---

the user, before the user starts typing. The user can consult this list to explore the vocabulary and guide their decision for the next word to write. The ranking of the items in the list is important, with more strongly endorsed words appearing higher up. Too many displayed options can slow down skilled users (Langlais and Lapalme, 2002), therefore the list should not be too long.

Typically, a language model is used to estimate the probability of the next word w_t given the typed word history $w_1, .., w_{t-1}$ and external context. The N-best list of the words with the highest probability is presented as the suggestions.

Word completion is a more interactive task, where the system makes suggestions to complete the current word as the user types each character. Here, the user has a clear intention of typing a specific word and the system should help them achieve this as quickly as possible. A single suggestion is presented and the user can choose to complete the word, typically by typing a special character (e.g. tab).

Word completion is based on interactive prefix matching against a lexicon, as shown in Algorithm 1. The algorithm takes as input the set of known vocabulary words and a scoring function that returns the goodness of a word in the current position and context, which again can be the word probability from a language model. Initialisation sets the prefix to an empty string and the lexicon to the whole vocabulary (lines 2-3). Iteratively, words that do not match with the prefix are removed from the lexicon (line 5), the best word from the lexicon according to the scorer is found and displayed to the user (lines 6-7) and the user can respond with a key (line 8). If the user inputs the special character, the best word is automatically completed (lines 9-10). If the user inputs a whitespace character, the algorithm terminates (11-12). This is the case when no matching word is found in the vocabulary. If any other character is typed, it is appended to the prefix and another iteration begins.

3.2 Neural language models

A language model (LM) estimates the probability of the next token given the previous tokens, i.e. $p(w_t|w_1, ..., w_{t-1})$. Recurrent neural networks (RNNs) have been successfully used for language modelling (Mikolov et al., 2010). Let w_t also denote the one-hot representation of the t-th token, i.e. w_t is a sparse binary vector with a single element set to 1, whose index uniquely identifies the token among a vocabulary of V known words. A neural LM uses a matrix, $E_{in} \in \mathbb{R}^{D \times V}$, to derive word embeddings, $e_t^w = E_{in}w_t$, where D is a latent dimension. A hidden state from the previous time step, h_{t-1}, and the current word embedding, e_t^w, are sequentially fed to an RNN's recurrence function to produce the current hidden state, $h_t \in \mathbb{R}^D$. The conditional probability of the next word is estimated as $\text{softmax}(E_{out}h_t)$, where $E_{out} \in \mathbb{R}^{V \times D}$ is an output embeddings matrix.

We use two extensions to the baseline neural LM, described in Spithourakis et al. (2016). A language model can be *conditioned* on the external context by using an encoder-decoder framework. The encoder builds a representation of the context, h_{KB}, which is then copied to the initial hidden state of the language model (decoder). To build such a representation for our structured context, we can lexicalise the KB by converting its tuples into textual statements of the form "*attribute : value*", which can then be encoded by an RNN. This approach can incorporate KB tuples flexibly, even when values of some attributes are missing.

The document and lexicalised KB will frequently contain numerical tokens, which are typically associated with high out-of-vocabulary rates. To make the LM more sensitive to such numerical information, we can define the inputs of the RNN's recurrence function at each time step as a concatenation of e_t^w and e_t^n, where the latter is a representation of the numeric value of w_t. We set $e_t^n = \text{float}(w_t)$, where $\text{float}(.)$ returns a floating point number from the string of its input or zero, if the conversion fails. When we train such a model, the representations for the words will be associated with the numerical values that appear in their context. Therefore, this model is numerically *grounded*.

3.3 Automated evaluation

We run an automated evaluation for both tasks and all systems by simulating a user who types the text character by character. The character stream comes from a dataset of finalised clinical reports. For the word prediction task, we assume that the word from the dataset is the correct word. For the word completion task, we assume that the user types the special key to autocomplete the word as soon as the correct suggestion becomes available.

In practice, the two tasks can be tackled at the same time, e.g. a list of suggestions based on a language model is shown as the user types and they can choose to complete the prefix with the word on the top of the list. However, we chose to decouple the two functions because of their conceptual differences, which call for different evaluation metrics.

For word prediction, the user has not yet started typing and they might seek guidance in the suggestions of the system for their final decision. A vocabulary exploration system will need to have a high recall. To also capture the effect of the length of the suggestions' list, we will report recall at various ranks (*Recall@k*), where the rank corresponds to the list length. Because our automated evaluation considers a single correct word, Recall@1 is numerically identical to Precision@1. We also report the mean reciprocal rank (*MRR*), which is the multiplicative inverse of the rank of the correct word in the suggestions' list. Finally, per token *perplexity* is

		train	dev	test
	#documents	11,158	1,625	3,220
	#KB tuples/doc	7.7	7.7	7.7
#tokens/ doc	all	204.9	204.4	202.2
	words	95.7%	95.7%	95.7%
	numeric	4.3%	4.3%	4.3%
OOV rate	all	5.0%	5.1%	5.2%
	words	3.4%	3.5%	3.5%
	numeric	40.4%	40.8%	41.8%
	#chars/token	4.9	4.9	4.9

Table 1: Statistics for clinical dataset. Counts for non-numeric (*words*) and *numeric* tokens reported as percentage of counts for *all* tokens. Out-of-vocabulary (OOV) rates are for vocabulary of 1000 most frequent words in the train data.

a common evaluation metric for language models.

For word completion, the main goal of the system should be to reduce input time and effort for the intended word that is being typed by the user. *Keystroke savings* (KS) measures the percentage reduction in keys pressed compared to character-by-character text entry. Suggestions that are not taken by the user are a source of unnecessary distractions. We define an *unnecessary distractions* (UD) metric as average number of unaccepted character suggestions that the user has to scan before completing a word.

$$KS = \frac{keys_{unaided} - keys_{with\ prediction}}{keys_{unaided}} \quad (1)$$

$$UD = \frac{count(suggested,\ not\ accepted)}{count(accepted)} \quad (2)$$

Bickel et al. (2005) note that KS corresponds to a recall metric and UD to a precision metric. Thus, we can use the F1 score (harmonic mean of precision and recall) to summarise both metrics.

$$Precision = \frac{count(accepted)}{count(suggested)} \quad (3)$$

$$Recall = \frac{count(accepted)}{count(total\ characters)} \quad (4)$$

4 Data

Our dataset comprises 16,003 anonymised clinical records from the London Chest Hospital. Table 1 summarises descriptive statistics of the dataset.

Each patient record consists of a text report and accompanying structured KB tuples. The latter describe metadata about the patient (age and gender) and results of medical tests (e.g. end diastolic and systolic volumes for the left and right ventricles as measured through magnetic resonance imaging). This information was extracted from the electronic health records held by the hospital and was available to the clinician at the time of the compilation of the report. In total, the KB describes 20 possible attributes. From these, one is categorical (gender) and the rest are numerical (age is integer and test results are real valued). On average, 7.7 tuples are completed per record.

Numeric tokens account for a large part of the vocabulary (>40%) and suffer from high out-of-vocabulary rates (>40%), despite constituting only a small proportion of each sentence (4.3%).

5 Results and discussion

In this section we describe the setup of our experiments (Subsection 5.1) and then present and discuss evaluation results for the word prediction (Subsection 5.2) and word completion (Subsection 5.3) tasks. Finally, we perform a qualitative evaluation (Subsection 5.4).

5.1 Setup

Our *baseline* LM is a single-layer long short-term memory network (LSTM) (Hochreiter and Schmidhuber, 1997) with all latent dimensions (internal matrices, input and output embeddings) set to $D = 50$. We extend this baseline model using the techniques described in Section 3.2 and derive a model conditional on the KB (+c), a model that is numerically grounded (+g) and a model that is both conditional and grounded (+c+g). We also experiment with ablations of these models that at test time ignore some source of information. In particular, we run the conditional models without the encoder, which ignores the KB (-kb), and the grounded models without the numeric representations, which ignores the magnitudes of the numerical values (-v).

	model	PP	MRR	Recall@1	Recall@3	Recall@5	Recall@10
system	baseline	14.96	17.19	8.36	18.38	25.03	36.66
system	+c	14.52	54.49	45.27	59.97	65.18	71.18
system	+g	9.91	31.91	21.13	35.45	43.66	53.72
system	+c+g	**9.39**	**60.71**	**51.76**	**66.36**	**71.28**	**77.10**
ablation	+c -kb	16.64	52.54	43.07	57.89	63.66	70.45
ablation	+g -v	13.16	56.08	46.58	61.96	67.30	73.49
ablation	+c+g -kb	10.82	58.72	49.46	64.31	69.71	75.98
ablation	+c+g -v	11.84	57.31	47.52	63.47	68.92	75.30
ablation	+c+g -kb-v	11.81	56.61	46.68	62.78	68.48	74.87

Table 2: Word-level evaluation results for next word prediction on the test set. Perplexity (PP), mean reciprocal rank (MRR) and Recall at different ranks. Recall@1 is equivalent to Precision@1. Best system values in **bold**.

	model	P	UD	KS(R)	F1
bound	theoretical	100.0	0.00	58.87	74.11
bound	vocabulary	100.0	0.00	54.48	70.54
system	baseline	13.96	6.17	34.35	19.85
system	+c	24.34	3.11	43.17	31.13
system	+g	18.60	4.38	39.31	25.25
system	+g+c	**26.60**	**2.76**	**44.81**	**33.38**
ablation	+c -kb	24.61	3.06	44.22	31.62
ablation	+g -v	26.74	2.74	45.71	33.74
ablation	+c+g -kb	26.73	2.74	45.72	33.74
ablation	+c+g -v	27.01	2.70	45.86	33.99
ablation	+c+g -kb-v	26.90	2.72	45.79	33.89

Table 3: Character-level evaluation results for word completion on the test set. Unnecessary distractions (UD) is inversely related to precision (P). Keystroke savings (KS) are equivalent with recall (R). Best system values in **bold**.

The vocabulary contains the $V = 1000$ most frequent tokens in the training set. Out-of-vocabulary tokens are substituted with $<num>$, if numeric, and $<unk>$, otherwise. We note that the numerical representations are extracted before any masking. Models are trained to minimise a cross-entropy loss, with 20 epochs of back-propagation and gradient descent with adaptive learing rates (AdaDelta) (Zeiler, 2012) and minibatch size set to 64. Hyperparameters are based on a small search on the development set around values commonly used in the literature.

5.2 Word prediction

We show our evaluation results on the test set for the word prediction task in Table 2. The conditioned model (+c) achieves double the MRR and quadruple the Recall@1 of the baseline model, despite bringing only small improvements in perplexity. The grounded model (+g) achieves a more significant perplexity improvement (33%), but smaller gains for MRR and Recall@1 (85% and 150% improvement, respectively). Contrary to intuition, we observe that a model with higher perplexity performs better in a language modelling task.

The grounded conditional model (+c+g) has the best performance among the systems, with about 5 points additive improvement across all evaluation metrics over the second best. The benefits from conditioning and grounding seem to be orthogonal to one another.

Recall increases with the length of the suggestion list (equivalent to rank). The increase is almost linear for the baseline, but for the grounded conditional it has a decreasing rate. The Recall@5 for the best model is similar to Recall@10 for the second best, thus allowing for halving the suggestions at the same level of recall.

In the test time ablation experiments, all evaluation metrics become slightly worse with the notable exception of the grounded without numerical values (+g-v), for which MRR and recall at all ranks are dramatically increased. Again, we observe that a worse perplexity does not always correlate with decreased performance for the rest of the metrics.

5.3 Word completion

We show our evaluation results on the test set for the word prediction completion in Table 3. In order to give some perspective to the results, we also compute upper bounds originally used to frame

document	system:	baseline	+c	+g	+c+g
left ventricular function		**suggestions**			
analysis results end — rank 1	1	non	normal	normal	preserved
diastolic volume <num>	2	normal	preserved	dilated	normal
ml end systolic volume	3	dilated	non	not	dilated
<num> ml stroke	4	preserved	good	preserved	not
volume <num> ml	5	not	mild	non	with
ejection fraction <num>		**ranks**			
% [...] lv systolic	non-dilated	10	11	8	13
function is moderately	dilated	3	8	2	3
impaired . non dilated	non	1	3	5	7
atria. non dilated rv [...]	moderately	41	33	37	36
lv is <word> dilated.	mildly	6	6	7	6
[...]	severely	29	23	28	27

(column 2 is labelled "rank" for rows 1–5 and "suggestion" for the rows below)

Table 4: Word prediction for sample document from the development set. Top-5 suggestion lists for <word> (original document has "non") and ranks for interesting terms from the complete lists of different systems.

<word>	numerical configuration		
	non	mild	severe
non	*85.83*	**50.45**	26.81
mildly	11.99	*36.27*	**46.46**
severely	2.18	13.28	*26.73*

Table 5: Document probabilities for different <word> choices and different numerical configurations. The probabilities are renormalised over the three displayed choices. Probabilities for highest scoring word in **bold** and for correct word in *italics*.

keystroke savings (Trnka and McCoy, 2008). The *theoretical* bound comes from an ideal system that retrieves the correct word after the user inputs the only the first character. The *vocabulary* bound is similar but only makes any suggestion if the correct word is in the known vocabulary. We extend these bounds to the rest of the evaluation metrics.

The conditioned model (+c) improves the keystroke savings by 25% over the baseline, while halving the unnecessary distractions. The grounded model (+g) achieves smaller improvements over the baseline. The grounded conditional model (+c+g) again has the best performance among the systems. It yields keystroke savings of 44.81%, almost halfway to the theoretical bound, and the lowest number of unnecessary distractions.

For this task, the desired behaviour of a system is to increase the keystroke savings without introducing too many unnecessary distractions (as measured by the number of wrongly suggested characters per word). Since the two quantities represent recall and precision measurements, respectively, a trade-off is expected between them (Bickel et al., 2005). Our extended models manage to improve both quantities without trading one for the other.

The theoretical and vocabulary bounds represent ideal systems that always make correct suggestions (UD=0). This translates into very high precision (100%) and F1 values (>70%) that purely represent upper bounds on these performance metrics. For reference, a system with the same keystroke savings as the theoretical bound (58.87%) and a single unnecessary character per word (UD=1) would achieve precision of 50% and an F1 score of 54.07%.

In the test time ablation experiments, all evaluation metrics have slightly better results than their corresponding system. In fact, some models perform similarly to the best system, if not marginally better.

5.4 Qualitative results

The previous results revealed two unexpected situations. First, we observed that occasionally a model with worse perplexity fares better at word prediction, which is a language modelling task. Second, we observed that occasionally a run time ablation of a conditional or grounded model outperforms its system counterpart. We carried out qualitative experiments in order to investigate these scenarios.

We selected a document from the development

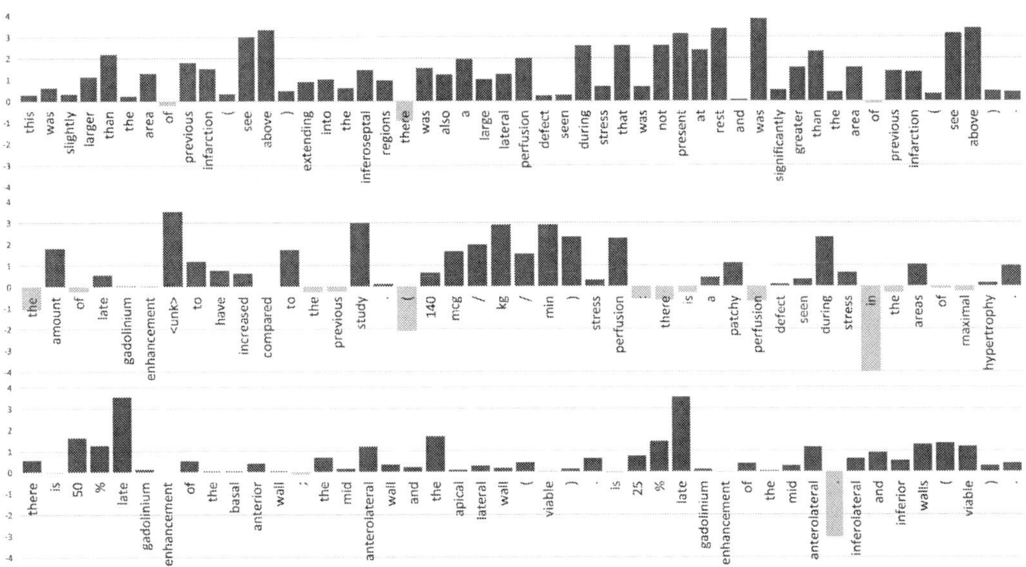

Figure 2: Word likelihood ratios (grounded conditional to baseline) for sample sentences from the development set.

set and identified a word of interest and numeric values that can influence the user's choice for that word. In Table 4, we show the selected document and the 5 top suggestions for the word by different systems. The systems do not have access to tokens from <word> onwards. We also show the ranks for several other semantically relevant choices that appear deeper in the suggestion list. Grounding and conditioning change the order in which the suggestions appear.

We proceeded to substitute the numeric values to more representative configurations that would each favour a particular word choice from the set {"non", "mildly", "severely"}. We found that changing the values does not have a significant effect to the suggestion probabilities and causes no reordering of the items in the lists shown in Table 4. This is in agreement with our previous results for test time ablations and can be attributed to the fact that many more parameters have been used to model words than numerical values. Thus, the systems rely less on numerical information at test time, even though at training time it helps to improve the language models.

Next, for the different numeric configurations we set <word> to each of the three choices and computed the probability of observing the whole document under the grounded model. This is done by multiplying together the probabilities for all individual words. Table 5 shows the resulting document probabilities, re-normalised over the three choices. We observe that the system has a stronger preference to "non", which happens to be the majority class in the training data. In contrast to word probabilities, document probabilities are influenced by the numerical configuration.

The reason for this difference in sensitivities is that the tiny changes in individual word probabilities accumulate multiplicatively to bring on significant changes in the document probability. Additionally, selecting a particular word influences the probabilities of the following words differently, depending on the numerical configuration. This also explains the observed differences between the perplexity of ablated systems, which accumulates small changes over the whole corpus, and the rest of the metrics, which only depend on per word suggestions. Our training objective, cross-entropy, is directly related to perplexity. Through this, numerical values seem to mediate at training time to learn a better language model.

Finally, we directly compare the word probabilities from different systems on several documents from the development set. In Figure 2 we plot the word likelihood ratio of the grounded conditional to baseline language models for three sentences. We can interpret the values on the vertical axis as how many times the word is more likely under the extended model versus the baseline. The probability of most words was increased, even at longer distances from numbers (first example). This is reflected in the improved perplexity of the language model. Words and contingent spans directly associated with numbers, such as units of measurement and certain symbols, also receive a boost (second example). Finally, the system would often recognise and penalise mistakes because of their unexpectedness (dot instead of a comma in the last example).

6 Conclusion

In this paper we showed how numerically grounded language models conditioned on an external knowledge base can be used in the tasks of word prediction and completion. Our experiments on a clinical dataset showed that the two extensions to standard language models have complimentary benefits. Our best model uses a combination of conditioning and grounding to improve recall from 25.03% to 71.28% for the word prediction task. In the word completion task, it improves keystroke savings from 34.35% to 44.81%, where the upper theoretical bound is 58.78% for this dataset. We found that perplexity does not always correlate with system performance in the two downstream tasks. Our ablation experiments and qualitative investigations showed that at test time numbers have more influence on the document level than on individual word probabilities.

Our approach did not rely on ontologies or fine grained data linkage. Such additional information might lead to further improvements, but would limit the ability of our models to generalise in new settings. While our automated evaluation showed that our extended system achieves notable improvements in keystroke savings, a case study would be required to measure the acceptance of such a system and its impact on clinical documentation processes and patient care. In the past, deployment of text prediction systems in clinical settings has lead to measurable gains in productivity (Hua et al., 2014; Gong et al., 2016).

In the future, we will investigate alternative ways to encode numerical information, in an attempt to improve the utilisation of numerical values at test time. We will also experiment with multitask objectives that consider numerical targets.

Acknowledgments

The authors would like to thank the anonymous reviewers. This research was supported by the Farr Institute of Health Informatics Research and an Allen Distinguished Investigator award.

References

Holger Bast and Ingmar Weber. 2006. Type less, find more: fast autocompletion search with a succinct index. In *Proceedings of the 29th annual international ACM SIGIR conference on Research and development in information retrieval*.

David Beukelman and Pat Mirenda. 2005. *Augmentative and alternative communication*. Brookes.

Steffen Bickel, Peter Haider, and Tobias Scheffer. 2005. Predicting sentences using n-gram language models. In *Proceedings of Human Language Technology and Empirical Methods in Natural Language Processing*.

Elia Bruni, Nam-Khanh Tran, and Marco Baroni. 2014. Multimodal distributional semantics. *Journal of Artificial Intelligence Research*, 49:1–47.

Mario Cannataro, Orlando Alfieri, and Francesco Fera. 2012. Knowledge-based compilation of magnetic resonance diagnosis reports in neuroradiology. In *25th International Symposium on Computer-Based Medical Systems (CBMS)*. IEEE.

Chi-Huang Chen, Sung-Huai Hsieh, Yu-Shuan Su, Kai-Ping Hsu, Hsiu-Hui Lee, and Feipei Lai. 2012. Design and implementation of web-based discharge summary note based on service-oriented architecture. *Journal of medical systems*, 36(1):335–345.

Kyunghyun Cho, Bart Van Merriënboer, Caglar Gulcehre, Dzmitry Bahdanau, Fethi Bougares, Holger Schwenk, and Yoshua Bengio. 2014. Learning phrase representations using rnn encoder-decoder for statistical machine translation. In *Proceedings of EMNLP*.

Marie-Catherine De Marneffe, Anna N Rafferty, and Christopher D Manning. 2008. Finding Contradictions in Text. In *Proceedings of ACL*.

Jeffrey Donahue, Lisa Anne Hendricks, Sergio Guadarrama, Marcus Rohrbach, Subhashini Venugopalan, Kate Saenko, and Trevor Darrell. 2015. Long-term recurrent convolutional networks for visual recognition and description. In *Proceedings of the IEEE Conference on Computer Vision and Pattern Recognition*.

Mark D Dunlop and Andrew Crossan. 2000. Predictive text entry methods for mobile phones. *Personal Technologies*, 4(2-3):134–143.

John Eng and Jason M Eisner. 2004. Informatics in radiology (inforad) radiology report entry with automatic phrase completion driven by language modeling. *Radiographics*, 24(5):1493–1501.

Afsaneh Fazly and Graeme Hirst. 2003. Testing the efficacy of part-of-speech information in word completion. In *Proceedings of the EACL 2003 Workshop on Language Modeling for Text Entry Methods*.

Michael Fleischman and Deb Roy. 2008. Grounded Language Modeling for Automatic Speech Recognition of Sports Video. In *Proceedings of ACL*.

George Foster, Philippe Langlais, and Guy Lapalme. 2002. User-friendly text prediction for translators. In *Proceedings of the ACL-02 conference on Empirical methods in natural language*.

Nestor Garay-Vitoria and Julio Abascal. 2006. Text prediction systems: a survey. *Universal Access in the Information Society*, 4(3):188–203.

Shalini Ghosh, Oriol Vinyals, Brian Strope, Scott Roy, Tom Dean, and Larry Heck. 2016. Contextual lstm (clstm) models for large scale nlp tasks. *arXiv:1602.06291*.

Yang Gong, Lei Hua, and Shen Wang. 2016. Leveraging user's performance in reporting patient safety events by utilizing text prediction in narrative data entry. *Computer methods and programs in biomedicine*, 131:181–189.

Sepp Hochreiter and Jürgen Schmidhuber. 1997. Long short-term memory. *Neural computation*, 9(8):1735–1780.

L Hua, S Wang, and Y Gong. 2014. Text prediction on structured data entry in healthcare: A two-group randomized usability study measuring the prediction impact on user performance. *Applied Clinical Informatics*, 5:249–263.

Douwe Kiela and Stephen Clark. 2015. Multi- and Cross-Modal Semantics Beyond Vision: Grounding in Auditory Perception. In *Proceedings of EMNLP*.

Douwe Kiela, Luana Bulat, and Stephen Clark. 2015. Grounding Semantics in Olfactory Perception. In *Proceedings of ACL*.

Philippe Langlais and Guy Lapalme. 2002. Trans type: Development-evaluation cycles to boost translator's productivity. *Machine Translation*, 17(2):77–98.

Ching-Heng Lin, Nai-Yuan Wu, Wei-Shao Lai, and Der-Ming Liou. 2014. Comparison of a semi-automatic annotation tool and a natural language processing application for the generation of clinical statement entries. *Journal of the American Medical Informatics Association*, 22:132–142.

Christian Lovis, Robert H Baud, and Pierre Planche. 2000. Power of expression in the electronic patient record: structured data or narrative text? *International Journal of Medical Informatics*, 58:101–110.

Bill MacCartney and Christopher D Manning. 2008. Modeling semantic containment and exclusion in natural language inference. In *Proceedings of the 22nd International Conference on Computational Linguistics*.

Tomáš Mikolov, Martin Karafiát, Lukáš Burget, Jan Černocký, and Sanjeev Khudanpur. 2010. Recurrent neural network based language model. In *Interspeech*.

Michael A Nakao and Seymour Axelrod. 1983. Numbers are better than words: Verbal specifications of frequency have no place in medicine. *The American Journal of Medicine*, 74(6):1061–1065.

Subhro Roy, Tim Vieira, and Dan Roth. 2015. Reasoning about Quantities in Natural Language. *Transactions of the Association for Computational Linguistics*, 3:1–13.

Mark Sammons, VG Vydiswaran, and Dan Roth. 2010. Ask not what textual entailment can do for you... In *Proceedings of the 48th Annual Meeting of the Association for Computational Linguistics*.

Merlijn Sevenster and Zharko Aleksovski. 2010. Snomed ct saves keystrokes: quantifying semantic autocompletion. In *Proceedings of American Medical Informatics Association (AMIA) Annual Symposium*.

Merlijn Sevenster, Rob van Ommering, and Yuechen Qian. 2012. Algorithmic and user study of an autocompletion algorithm on a large medical vocabulary. *Journal of biomedical informatics*, 45(1):107–119.

Carina Silberer and Mirella Lapata. 2014. Learning Grounded Meaning Representations with Autoencoders. In *Proceedings of ACL*.

Raul Sirel. 2012. Dynamic user interfaces for synchronous encoding and linguistic uniforming of textual clinical data. In *Human Language Technologies– The Baltic Perspective: Proceedings of the 5th International Conference Baltic HLT*.

Georgios P. Spithourakis, Isabelle Augenstein, and Sebastian Riedel. 2016. Numerically grounded language models for semantic error correction. In *Proceedings of EMNLP*.

Daniëlle Timmermans. 1994. The roles of experience and domain of expertise in using numerical and verbal probability terms in medical decisions. *Medical Decision Making*, 14(2):146–156.

Keith Trnka and Kathleen F McCoy. 2008. Evaluating word prediction: framing keystroke savings. In *Proceedings of the 46th Annual Meeting of the Association for Computational Linguistics on Human Language Technologies*.

Keith Trnka. 2008. Adaptive language modeling for word prediction. In *Proceedings of the 46th Annual Meeting of the Association for Computational Linguistics on Human Language Technologies*.

Antal Van Den Bosch and Toine Bogers. 2008. Efficient context-sensitive word completion for mobile devices. In *Proceedings of the 10th international conference on Human computer interaction with mobile devices and services.*

Oriol Vinyals, Alexander Toshev, Samy Bengio, and Dumitru Erhan. 2015. Show and tell: A neural image caption generator. In *Proceedings of the IEEE Conference on Computer Vision and Pattern Recognition.*

Tonio Wandmacher and Jean-Yves Antoine. 2008. Methods to integrate a language model with semantic information for a word prediction component. In *Proceedings of EMNLP.*

Li Yao, Atousa Torabi, Kyunghyun Cho, Nicolas Ballas, Christopher Pal, Hugo Larochelle, and Aaron Courville. 2015. Describing videos by exploiting temporal structure. In *Proceedings of the IEEE International Conference on Computer Vision.*

Matthew D. Zeiler. 2012. ADADELTA: An Adaptive Learning Rate Method. *CoRR*, abs/1212.5701.

Modelling Radiological Language with Bidirectional Long Short-Term Memory Networks

Savelie Cornegruta, Robert Bakewell, Samuel Withey and **Giovanni Montana**
Department of Biomedical Engineering, King's College London, UK
`giovanni.montana@kcl.ac.uk`

Abstract

Motivated by the need to automate medical information extraction from free-text radiological reports, we present a bi-directional long short-term memory (BiLSTM) neural network architecture for modelling radiological language. The model has been used to address two NLP tasks: medical named-entity recognition (NER) and negation detection. We investigate whether learning several types of word embeddings improves BiLSTM's performance on those tasks. Using a large dataset of chest x-ray reports, we compare the proposed model to a baseline dictionary-based NER system and a negation detection system that leverages the hand-crafted rules of the NegEx algorithm and the grammatical relations obtained from the Stanford Dependency Parser. Compared to these more traditional rule-based systems, we argue that BiLSTM offers a strong alternative for both our tasks.

1 Introduction

Radiological reports represent a large part of all Electronic Medical Records (EMRs) held by medical institutions. For instance, in England alone, upwards of 22 million plain radiographs were reported over the 12-month period from March 2015 (NHS, 2016). A radiological report is a written document produced by a Radiologist, a physician that specialises in interpreting medical images. A report typically states any technical factors relevant to the acquired image as well as the presence or absence of radiological abnormalities. When an abnormality is noted, the Radiologist often gives further description, including anatomical location and the extent of the disease.

Whilst Radiologists are taught to review radiographs in a systematic and comprehensive manner, their reporting style can vary quite dramatically (Reiner and Siegel, 2006) and the same findings can often be described in a multitude of different ways (Sobel et al., 1996). The radiological reports may contain broken grammar and misspellings, which are often the result of voice recognition software or the dictation-transcript method (McGurk et al., 2014). Applying text mining techniques to these reports poses a number of challenges due to extensive variability in language, ambiguity and uncertainty, which are typical problems for natural language.

In this work we are motivated by the need to automatically extract standardised clinical information from digitised radiological reports. A system for the fully-automated extraction of this information could be used, for instance, to characterise the patient population and help health professionals improve day-to-day services. The extracted structured data could also be used to build management dashboards (Simpao et al., 2014) summarising and presenting the most prevalent conditions. Another potential use is the automatic labelling of medical images, e.g. to support the development of computer-aided diagnosis software (Shin et al., 2015).

In this paper we propose a recurrent neural network (RNN) architecture for modelling radiological language and investigate its potential advantages on two different tasks: medical named-entity recognition (NER) and negation detection. The model, a bi-directional long short-term memory (BiLSTM) network, does not use any hand-engineered features,

Proceedings of the Seventh International Workshop on Health Text Mining and Information Analysis (LOUHI), pages 17–27,
Austin, TX, November 5, 2016. ©2016 Association for Computational Linguistics

but learns them using a relatively small amount of labelled data and a larger but unlabelled corpus of radiological reports. In addition, we explore the combined use of BiLSTM with other language models such as GloVe (Pennington et al., 2014) and a novel variant of GloVe, proposed here, that makes use of a medical ontology. The performance of the BiLSTM model is assessed comparatively to a rule-based system that has been optimised for the tasks at hand and builds upon well established techniques for medical NER and negation detection. In particular, for NER, the system uses a baseline dictionary-based text mining component relying on a curated dictionary of medical terms. As a baseline for the negation detection task, the system implements a hybrid component based on the NegEx algorithm (Chapman et al., 2013) in conjunction with grammatical relations obtained from the Stanford Dependency Parser (Chen and Manning, 2014).

The article is organised as follows. In Section 2 we provide a brief review of the existing body of work in NLP for medical information extraction and briefly discuss the use of artificial neural networks for NLP tasks. In Section 3 we describe the datasets used for our experiments, and in Section 4 we introduce the BiLSTM model. The results are presented in Section 6 where we also compare BiLSTM against the rule-based baseline systems described in Section 5.

2 Related Work

2.1 Medical NER

A large proportion of NLP systems for medical text mining use dictionary-based methods for extracting medical concepts from clinical document (Friedman et al., 1995; Johnson et al., 1997; Aronson, 2001; Savova et al., 2010). The dictionaries that contain the correspondence between a single- or multi-word phrase and a medical concept are usually built from medical ontologies such as the Unified Medical Language System (UMLS) (NLM, 2016b) and Medical Subject Headings (MeSH) (NLM, 2016a). These ontologies contain hundreds of thousands of medical concepts. There are also domain-specific ontologies such as RadLex (Langlotz, 2006), which has been developed for the Radiology domain, and currently contains over $68,000$ concepts.

Medical Language Extraction and Encoding System (MEDLEE) (Friedman et al., 1995) is one of the earliest automated systems originally developed for handling radiological reports, and later expanded to other medical domains. MEDLEE parses the given clinical documents by string matching: the words are matched to a pre-defined dictionary of medical terms or semantic groups (e.g. *Central Finding*, *Bodyloc Modifier*, *Certainty Modifier* and *Region Modifier*). Once the words have been associated with a semantic group, a Compositional Regularizer stage combines them according to a list of pre-defined mappings to form regularized multiword phrases. The final stage looks up the regularized terms in a dictionary of medical concepts (e.g. *enlarged heart* is mapped to the corresponding concept *cardiomegaly*). A separate study evaluated MEDLEE on 150 manually annotated radiology reports (Hripcsak et al., 2002); MEDLEE was assessed on its ability to detect 24 clinical conditions achieving an average sensitivity and specificity of 0.81 and 0.99, respectively.

A more recent system for general medical information extraction is the Mayo Clinic's Text Analysis and Knowledge Extraction System (cTAKES) (Savova et al., 2010), which also implements an NLP pipeline. During an initial shallow parsing stage, cTAKES attempts to group words into multiword expressions by identifying constituent parts of the sentence (e.g. noun, prepositional, and verb phrases). It then string matches the identified phrases to a concept in UMLS. A new set of semantic groups were also derived from the UMLS ontology (Ogren et al., 2007). The NER performance of the cTAKES was evaluated on the semantic groups, achieving an F1-score of 0.715 for exact matches and 0.824 for overlapping matches.

In general, dictionary-based systems perform with high precision on the NER tasks but have a low recall, showing a lack of generalisation. Low recall is usually caused by the inability to identify multiword phrases as concepts, unless exact matches can be found in the dictionary. In addition, such systems are not able to easily deal with disjoint entities. For instance, in the phrase *lungs are mildly hyperexpanded*, *hyperexpanded lungs* constitutes a clinical finding. In an attempt to deal with disjoint entities, rule-based systems such as MEDLEE,

MetaMap (Aronson, 2001) and cTAKES, implement additional parsing stages to find grammatical relations between different words in a sentence, thus aiming to create disjoint multi-word phrases. However, state-of-the-art syntactic parsers are still likely to fail when parsing sentences with broken grammar, as often occurs in clinical documents.

In an attempt to improve upon dictionary-based information extraction systems, Hassanpour (2015) recently used a first-order linear-chain Conditional Random Field (CRF) model (Lafferty et al., 2001) in a medical NER task involving five semantic groups (anatomy, anatomy modifier, observation, observation modifier, and uncertainty). The features used for the CRF model included part-of-speech (POS) tags, word stems, word n-grams, word shape, and negations extracted using the NegEx algorithm. The model was trained and tested using 10-fold cross validation on a corpus of 150 multi-institutional Radiology reports and achieved a precision score of 0.87, recall of 0.84, and F1-score of 0.85.

2.2 Medical negation detection

NegEx, a popular negation detection algorithm, is usually applied to medical concepts after the entity recognition stage. This tool uses a curated list of phrases (e.g. *no, no sign of, free of*), which are string matched to the medical text to detect a negation trigger, i.e. a word or phrase indicating the presence of a negated medical entity in the sentence. The target entities falling inside a window, starting at the negation trigger, are then classified as *negated*. In light of its simplicity, speed and reasonable results, NegEx had been used as a component by many medical NLP systems (Wu et al., 2014). It has been shown that that NegEx achieves an accuracy of 0.94 as part of the cTAKES evaluation (Savova et al., 2010). However, the window approach that is used for classifying the negations may result in a large number of false positives, especially if there are multiple entities within the 6-word window.

Aiming to reduce the number of false positives, recent efforts have integrated NegEx with machine learning models that can be trained on annotated datasets. For instance, Shivade (2015) introduced a kernel-based approach that uses features built using the type of negation trigger, features that are derived from the existence of conjunctions in the sentence, and features that weight the NegEx output against the bag-of-words in the dataset. The kernel based model outperformed the original NegEx algorithm by 2.7 F1-score points when trained and tested on the NegEx dataset. At around the same time, Mehrabi (2015) introduced DEEPEN, an algorithm that filters the NegEx output using the grammatical relations extracted using Stanford Dependency Parser. DEEPEN succeeded at reducing the number of false positives, although it showed a marginally lower F1-score when compared with NegEx on concepts from the *Disorders* semantic group from the Mayo Clinic dataset (Ogren et al., 2007).

2.3 Neural networks for NLP tasks

In recent years, deep artificial neural networks have been found to yield consistently good results on various NLP tasks. The SENNA system (Collobert et al., 2011), which used a convolutional neural network (CNN) architecture, came close to achieving state-of-the-art performance across the tasks of POS tagging, shallow parsing, NER, and semantic role labeling. More recently, recurrent neural networks (RNNs) have been shown to achieve very high performance, and often reach state-of-the-art results in various language modelling tasks (Mikolov and Zweig, 2012). RNNs have also been shown to outperform more traditional machine learning models, such as Logistic Regression and CRF, at the slot filling task in spoken language understanding (Mesnil et al., 2013). In a NER task on the publicly available datasets in four languages, the bidirectional long short-term memory (LSTM) networks (Hochreiter and Schmidhuber, 1997), a variant of RNN, outperformed CNNs, CRFs and other models (Lample et al., 2016).

Neural networks have also been used to learn language models in an unsupervised learning setting. Some popular models include Skip-gram and continuous bag-of-words (CBOW) (Mikolov et al., 2013). These yield word representations, or embeddings, that are able to carry the syntactic and semantic information of a language. Collobert (2011) showed that integrating pre-trained word embeddings into a neural network can help the supervised learning process.

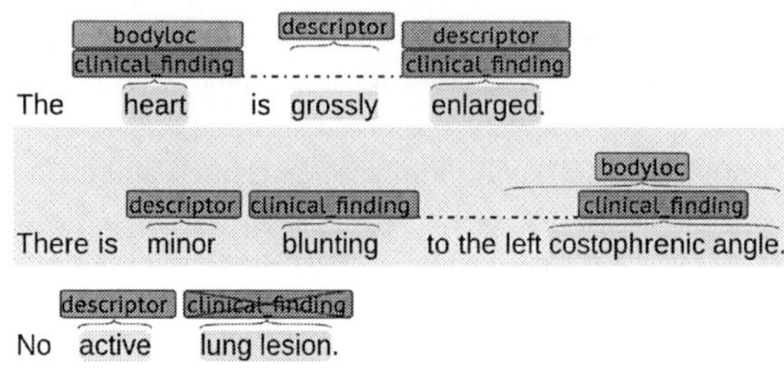

Figure 1: Example of manual annotation of a radiology report performed using BRAT

Semantic Group	# of entities	# of tokens
Body Location	5686	10113
Clinical Finding	5396	8906
Descriptor	3458	3845
Medical Device	1711	3361
Total	16251	26225
Negated entities	1851	2557

Table 1: Frequency distribution of entities by class in $2,000$ manually annotated reports

3 A Radiology corpus

3.1 Dataset

For this study, we produced an in-house radiology corpus consisting of $745,480$ historical chest X-ray (radiographs) reports provided by Guy's and St Thomas' Trust (GSTT). This Trust runs two hospitals within the National Health Service (NHS) in England, serving a large area in South London. The reports cover the period between January 2005 and March 2016, and were generated by 276 different reporters including consultant Radiologists, trainee Radiologists and reporting Radiographers. Our repository consists of text written or dictated by the clinicians after radiograph analysis, and do not contain any referral information or patient-identifying data, such as names, addresses or dates of birth. However, many reports refer to the clinical history of the patient. The reports had a minimum of 1 word and maximum of 311 words, with an average of 25.3 words and a standard deviation of 19.9 words. On average there were 2.9 sentences per report. After lemmatization, converting to lower case, and discounting words that occur less than 3 times in the corpus, the resulting vocabulary contained $8,031$ words.

A sample of $2,000$ reports was randomly selected from the corpus for the purpose of creating a training and validation dataset for the NER and negation detection tasks, whilst the remaining of the reports were utilised for pre-training word embeddings. The reports selected for manual annotation were written for all types of patients (Inpatient: 1072, A&E Attender: 515, Outpatient: 229, GP Direct Access Patient: 165, Ward Attender: 9, Day Case Patient: 8) by 144 different clinicians.

We introduce a simple word-level annotation schema that includes four classes or semantic groups: *Clinical Finding*, *Body Location*, *Descriptor* and *Medical Device*: Clinical Finding encompasses any clinically-relevant radiological abnormality, Body Location refers to the anatomical area where the finding is present, and Descriptor includes all adjectives used to describe the other classes. The Medical Device class is used to label any medical apparatus seen on chest radiographs, such as pacemakers, intravascular lines, and nasogastric tubes. Our annotation schema allows for the same token to belong to several semantic groups. For example, as shown in Figure 1, the word *heart* was associated with both Clinical Finding and Body Location classes. We have also introduced a negation attribute to indicate the absence of any of these entities.

3.2 Gold standard

Two clinicians (RB and SW) annotated the reports using BRAT (Stenetorp et al., 2012), a collaborative tool for text annotation that was configured to use our own schema. The BRAT output was then transformed to the IOBES tagging schema. Here, we interpret I as a token in the middle of an entity; O as a token not part of the entity; B and E as the beginning and end of the entity, respectively; finally, S indicates a single-word entity. We work with the assumption that entities may be disjoint and tokens that are surrounded by disjoint entity may belong to a different semantic group. For example, according to the annotation performed by the clinicians, in the sentence *Heart is slightly enlarged* the phrase *heart enlarged* represents an entity that belongs to the semantic group *Clinical Finding* and *slightly* is a *Descriptor*. The resulting breakdown of all entities by semantic group can be found in Table 1.

4 Methodology

In this Section we describe a model for NER that extracts five types of entities: the four semantic groups described in Section 3.1, as well as the negation, which is treated here as an additional class, analogously to the semantic groups.

4.1 Bi-directional LSTM

The RNN is a neural network architecture designed to model time series, but it can be applied to other types of sequential data (Rumelhart et al., 1988). As the information passes through the network, it can persist indefinitely in its memory. This facilitates the process of capturing sequential dependencies. The RNN makes a prediction after processing each element of the input sequence. Hence, the output sequence can be of the same length as the input sequence. The RNN architecture lends itself as a natural model for the proposed NER task, where the objective is to predict the IOBES tags for each of the input words.

The RNN is trained using the error backpropagation through time algorithm (Werbos, 1990) and a variant of the gradient descent algorithm. However, training these models is notoriously challenging due to the problem of exploding and vanishing gradients, especially when trained with long input sequences (Bengio et al., 1994). For the exploding gradient problem, numerical stability can be achieved by clipping the gradients (Graves, 2013). The problem of vanishing gradients can be addressed by replacing the standard RNN cell with a long short-term memory (LSTM) cell, which allows for a constant error flow along the input sequence (Hochreiter and Schmidhuber, 1997). A more constant error also means that the network is able to learn better long-term dependencies over the input sequence. By combining the outputs of two RNNs that pass the information in opposing directions, it is possible to capture the context from both ends of the sequence. The resulting architecture is known as Bidirectional LSTM (BiLSTM) (Graves and Schmidhuber, 2005).

We start by defining a vocabulary $V = \{v_1, v_2, ..., v_{8031}\}$ that contains the words extracted from the corpus as described in Section 3.1. We assume that, in order to perform NER on the words in any given sentence, it is sufficient to consider only the information contained in that sentence. Therefore we pass the BiLSTM one sentence at a time. For each input sentence of n words we define an n-dimensional vector $\mathbf{x}$ whose elements are the indices in V corresponding to words appearing in the sentence, preserving the order. The input $\mathbf{x}$ is passed to an Embedding Layer that returns the sequence $S = \{w_j | j = x_1, x_2, ..., x_n\}$ where w_j is the jth row of a dense matrix $\mathbf{W} \in \mathbb{R}^{|V| \times d}$, where $d \in \mathbb{N}$ is a hyperparameter. The vector w_j represents a low-dimensional vector representation, or word embedding, whereas $\mathbf{W}$ is the corresponding embedding matrix. The sequence of word embeddings S is then passed as input to two LSTM layers that process it in opposing directions (forwards and backwards), similar to the architecture introduced by Graves (2005). Figure 2 shows the LSTM layers in their "unrolled" form as they read the input. Each LSTM layer contains k LSTM memory cells which are based on the implementation by Graves (2013). The output from each of the LSTM layers is $H = \{\mathbf{h}_t \in \mathbb{R}^k | t = 1, 2, ..., n\}$.

Next, we concatenate and flatten $H_{forward}$ and $H_{backward}$, obtaining a vector $\mathbf{p} \in \mathbb{R}^{2kn}$. We pass $\mathbf{p}$ through a linear transformation layer and reshape its output to a tensor of size $n \times C \times T$, where C is the number of annotation classes (5 in total, 4 semantic groups and 1 class for negation) and T is the number of possible tags (5 for the IOBES tags). Finally we apply the softmax function along the last dimension of the tensor to approximate the probability for each of the possible tags for each of the annotation class.

4.2 Word embeddings

We explored 4 different techniques for learning word embeddings from the text. The embeddings will subsequently be used to initialise the embedding matrix $\mathbf{W}$ that is required by BiLSTM for the NER task. In previous work, the initialisation of $\mathbf{W}$ with pre-trained embeddings has been found to improve the training process (Collobert et al., 2011; Mesnil et al., 2013).

Random Embeddings

Random embeddings were obtained by drawing from a uniform distribution in the $(-0.01, 0.01)$ range. As such, the positions of the words in the vector space do not provide any information regard-

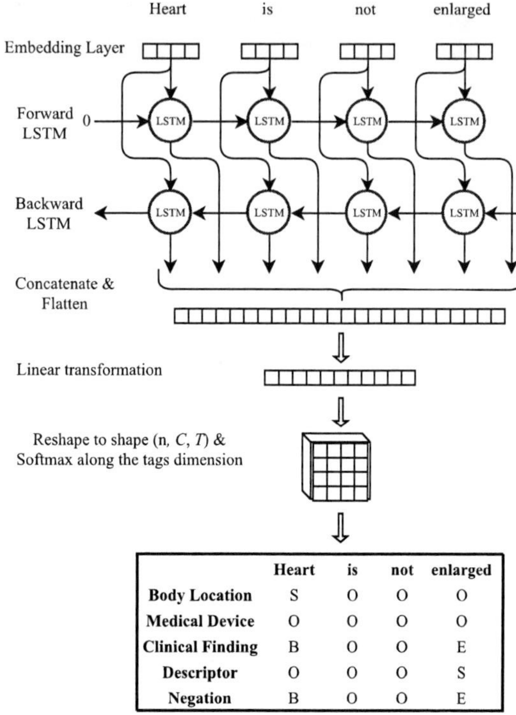

	Heart	is	not	enlarged
Body Location	S	O	O	O
Medical Device	O	O	O	O
Clinical Finding	B	O	O	E
Descriptor	O	O	O	S
Negation	B	O	O	E

Figure 2: An illustration of the BiLSTM architecture for joint medical entity recognition and negation detection

ing patterns of relationships between words.

BiLSTM Embeddings

These embeddings were obtained after adapting the BiLSTM for a language modelling task. Following a previously described strategy (Collobert and Weston, 2008), the input words were randomly replaced, with probability 0.2, with a word extracted from V. We then created a corresponding vector of binary labels to be used as prediction targets: each element of the vector is either 0 or 1, where 0 indicates a word that has been replaced, and 1 indicates an unchanged word. The model outputs the probability of the labels for each word in the given sentence. After training this language model on the unlabelled part of our corpus, we extracted the word embeddings from $\mathbf{W}$.

GloVe Embeddings

Word embedding were also obtained using GloVe, an unsupervised method (Pennington et al., 2014). On word similarity and analogy tasks, it has the potential to outperform competing models such as Skip-gram and CBOW. The GloVe objective function is

$$\sum_{i,j=1}^{|V|} f(X_{ij})(w_i^T \tilde{w} + b_i + \tilde{b}_j - log X_{ij})^2$$

where X is the word-word co-occurrence matrix, f is a weighting function, w are word embeddings, and $\tilde{w} \in \mathbb{R}^d$ are context word embeddings, with b and $\tilde{b}$ the respective bias terms. The GloVe embeddings w are trained using AdaGrad optimisation algorithm (Duchi et al., 2011), stochastically sampling nonzero elements from X.

GloVe-Ontology Embeddings

Furthermore, we introduced a modified version of GloVe, denoted GloVe-Ontology, with the objective to leverage the RadLex ontology during the word embedding estimation process. The rationale is to impose some constrains on the estimated distance between words using semantic relationships extracted from RadLex; this is an idea somewhat inspired by previous work (Yu and Dredze, 2014).

The RadLex data was initially represented as a tree, τ, by considering only the relation *is_parent_of* between concepts. We then attempted to string match every word v in V to a concept in τ. Every v matched with a RadLex concept was then assigned the vector that enumerates all ancestors of that concept; otherwise it was associated with a zero vector. We denote the resulting vector by ϕ. We imposed the constraint that words close to each other in τ should also be close in the learned embedding space. Accordingly, GloVe's original objective function was modified to incorporate this additional penalty:

$$\sum_{i,j=1}^{|V|} f(X_{ij})(w_i^T \tilde{w} + b_i + \tilde{b}_j - log X_{ij} - \alpha sim(\phi_i, \phi_j))^2$$

In this expression, α is a parameter controlling the influence of this additional constraint, and sim is taken to be the cosine similarity function. No major changes in the training algorithm were required compared to the original GloVe methodology.

4.3 BiLSTM implementation and training

The BiLSTM was implemented using two open-source libraries, *Theano* (Theano Development

Team, 2016) and *Lasagne* (Dieleman et al., 2015). The number of memory cells in each LSTM layer, k, was set to 100. We limited the maximum length of the input sequence to 40 words and for shorter inputs we used a binary mask at the input and cropped the output predictions accordingly. The loss function was the categorical cross-entropy between the predicted probabilities of the IOBES tags and the true tags. BiLSTM was trained on a GPU for 20 epochs in batches of 10 sentences using Stochastic Gradient Descent (SGD) with Nesterov momentum and with the learning rate set to 0.5.

The embedding size d was set to 50. The GloVe, GloVe-Ontology and BiLSTM word embeddings were trained on $743,480$ unlabelled radiology reports. The α paramenter in the Glove-Ontology objective was set to 0.5.

One aspect of the training was to allow or block the optimisation algorithm from updating the matrix $\mathbf{W}$ in the Embedding Layer of the BiLSTM. In Section 6 we refer to this aspect of training as *fine-tuning*. Previous work (Collobert et al., 2011) has shown that fine-tuning can boost the results of the several supervised tasks in NLP.

5 A competing rule-based system

Two clinicians (RB and SW) built a comprehensive dictionary of medical terms. In the dictionary, the key is the name of the term and the corresponding value specifies the semantic group, which was identified using a number of resources. We iterated over all RadLex concepts using the field *Preferred Label* as the dictionary key for the new entry. To obtain the semantic group we traversed up the ontology tree until an ancestor concept was found that had been manually mapped to a semantic group. For example, one of the ancestor concepts of *heart* is *Anatomical entity*, which we had manually mapped to semantic group *Body Location*. The same procedure was also performed on the MeSH ontology using the *MeSH Heading* field as a dictionary key. Finally, we added 202 more terms that were common in day-to-day reporting but were not present in RadLex and MeSH.

The sentences were tokenized and split using the Stanford CoreNLP suite (Manning et al., 2014), and also converted to lower case and lemmatized using NLTK (Bird et al., 2009). Next, for each sentence, the algorithm attempted to match the longest possible sequence of words, a target phrase, to an entry in the dictionary of medical terms. When the match was successful, the target phrase was annotated with the corresponding semantic group. When no match was found, the algorithm attempted to look up the target phrase in the English Wikipedia redirects database. In case of a match, the name of the target Wikipedia article was checked against our curated dictionary and the target phrase was annotated with the corresponding semantic group (e.g. *oedema* redirects to *edema*, which is how this concept is named in RadLex).

For all the string matching operations we used SimString (Okazaki and Tsujii, 2010), a fast and efficient approximate string matching tool. We arbitrarily chose the *cosine* similarity measure and a similarity threshold value of 0.85. Using SimString allowed the system to match misspelled words (e.g. *cardiomegally* to the correct concept *cardiomegaly*).

For negation detection, the system first obtained NegEx predictions for the entities extracted in the NER task. Next, it generated a graph of grammatical relations as defined by the Universal Dependencies (De Marneffe et al., 2014) from the Stanford Dependency Parser. It then removed all relations in the graph except *neg*, the negation relation, and *conj:or*, the *or* disjunction. Given the NegEx output and the reduced dependency graph, the system finally classified an entity as negated if any of the following two conditions were found to be true: (1) any of the words that are part of the entity were in a *neg* relation or in a *conj:or* relation with a another word that was in a *neg* relation; (2) if an entity was classified by NegEx as negated, it was the closest entity to negation trigger and there was no *neg* relations in the sentence. Our hybrid approach is somewhat similar to DEEPEN with the difference that the latter considers all first-order dependency relations between the negation trigger and the target entity.

6 Experimental Results

We evaluated the BiLSTM model on the medical NER task by measuring the overlap between the predicted semantic groups and the ground truth labels. The evaluation was performed at the granularity of a single word and using 5-fold cross-validation. The BiLSTM model was always trained on 80% of the annotated corpus and tested on the remaining 20%.

Embeddings	Fine-tuning	P	R	F1
Random	TRUE	0.878	0.869	0.873
Glove	TRUE	0.869	0.829	0.849
Glove-ontology	TRUE	0.875	0.860	0.867
BiLSTM	TRUE	0.878	0.870	**0.874**
Random	FALSE	0.829	0.727	0.775
Glove	FALSE	0.866	0.828	0.847
Glove-ontology	FALSE	0.850	0.839	0.844
BiLSTM	FALSE	0.870	0.849	**0.859**
Rule-based		0.706	0.698	0.702

Table 2: Comparison of the BiLSTM model and rule-based system. BiLSTM is trained using different word embedding and evaluated using 5-fold cross-validation. The evaluation considers the overlap span of the semantic group predictions against gold standard annotations.

Semantic Group	P	R	F1
Body Location	0.896	0.887	0.891
Medical Device	0.898	0.923	0.910
Clinical Finding	0.871	0.895	0.883
Descriptor	0.824	0.725	0.771
Total	0.878	0.870	0.874

Table 3: BiLSTM: performance metrics broken down by semantic group for the NER task. All results were obtained using BiLSTM word embeddings.

Semantic Group	P	R	F1
Body Location	0.724	0.839	0.778
Medical Device	0.976	0.538	0.694
Clinical Finding	0.862	0.551	0.672
Descriptor	0.467	0.780	0.584
Total	0.706	0.698	0.702

Table 4: Rule-based system: performance metrics broken down by by semantic group for the NER task.

Model	P	R	F1
BiLSTM	0.903	0.912	0.908
NegEx	0.664	0.944	0.780
NegEx - Stanford	0.944	0.912	**0.928**

Table 5: Comparison of BiLSTM, NegEx and NegEx-Stanford for negation detection. All algorithms predicted whether a given medical entity was negated or affirmed.

Table 2 compares the performance of various BiLSTM variants that were obtained with and without fine-tuning of the word embeddings to the perfor-

nodule	pacemaker	small	remains	fracture
bulla	ppm	tiny	remain	fractures
nodules	icd	minor	appears	deformity
opacity	wires	mild	is	body
opacities	drains	dense	are	scoliosis
opacification	leads	extensive	were	abnormality

Table 6: For each one of the five words in boldface, five nearest neighbours found in the embedding space learnt by BiLSTM.

mance of our baseline rule-based system. Without fine-tuning, the BiLSTM NER model, that was initialised with the embeddings trained in an unsupervised manner using the BiLSTM language model, achieves the best F1-score (0.859), and outperforms the next best variant by 0.012. With fine-tuning, the same BiLSTM variant improves the F1-score by a further 0.015 and outperforms the baseline rule-based system by an F1-score of 0.172. Table 3 shows its performance measure for each of the semantic groups.

The evaluation of negation detection was measured on complete entities. If any of the words within an entity were tagged with a I, B, E or S, that entity was considered to be negated. As shown in Table 5, the BiLSTM (BiLSTM language model embeddings, fine-tuning allowed) achieved an F1-score of 0.902, which outperformed NegEx by 0.128. However, the best F1-score of 0.928 is achieved using the NegEx-Stanford system.

7 Discussion

In Table 3, we show the predictive performance of the best BiLSTM NER model for each of the semantic groups. *Body Location*, *Medical Device* and *Clinical Finding* show a balanced precision and recall, and similar F1-scores. *Descriptor* has a lower F1-score which is caused by a low recall that may be the results of the larger variability in the words used for this semantic group. Table 4 shows the corresponding results for the rule-based NER system. *Medical Device* and *Clinical Finding* show a typical performance for a dictionary-based NER system with a high precision and a low recall. *Body Location* has relatively high precision and recall values which suggests that this semantic group is well covered by our dictionary of medical terms. In contrast, *Descriptor* shows a very low precision which is the result of a high number of false positives. The false positives are caused by many *Descriptor* entries in our dictionary of medical terms that had been automatically extracted from RadLex and MeSH but which do not correspond to the definition of a *Descriptor* used by the clinicians who produced the labelled data.

As a qualitative assessment, Table 6 shows the 5 nearest neighbours obtained from BiLSTM language model embeddings of some frequent words used by Radiologists. We note that there is an clear semantic similarity between the nearest neighbour words. Additionally, the embeddings encode syntactic information as the nearest neighbour words are parts of speech of the same type as the target word. We also summed the vectors for *heart* and *enlarged*, which yielded vec(*cardiomegaly*) as the nearest vector. Similarly, the closest vector to vec(*heart*) + vec(*not*) + vec(*enlarged*) is vec(*normal*). These examples suggest that word embeddings may encode information about the compositionality of words as discussed by Mikolov (2013).

Table 2 shows that, without fine-tuning, the Embedding Layer weights can affect the performance of the NER task. When fine-tuning is allowed there is only a marginal advantage in using pre-trained embeddings, as the BiLSTM performs equally well when initialised with random embeddings. Therefore, despite a positive qualitative assessment, the pre-trained word embeddings seem to offer only a small advantage when used for the proposed NER task as BiLSTM is able to learn well using the annotated data during the supervised learning phase.

8   Conclusions

In this paper we have shown that a recurrent neural network architecture, BiLSTM, can learn to detect clinical findings and negations using only a relatively small amount of manually labelled radiological reports. Using a manually curated medical corpus, we have provided initial evidence that BiLSTM outperforms a dictionary-based system on the NER task. For the detection of negations, on our dataset BiLSTM approaches the performance of a negation detection system that was build using the popular NegEx algorithm and uses grammatical relations obtained from the Stanford Dependency Parser and hand-crafted rules. We believe that increasing the size of the annotated training dataset can result in much improved performance on this task, and plan to purse this investigation in future work.

We have also investigated potential performance gains that can be achieved by using pre-trained word embeddings, i.e. BiLSTM, GloVe and GloVe-Ontology embeddings, in the context of BiLSTM-based modelling for the NER task. Our initial experimental results suggest that there is marginal benefit in using BiLSTM-learned embeddings while pre-training using GloVe and GloVe-Ontology embeddings did not offer any significant improvements over a random initialisation.

References

Alan R Aronson. 2001. Effective mapping of biomedical text to the UMLS Metathesaurus: the MetaMap program. In *Proceedings of the AMIA Symposium*, page 17. American Medical Informatics Association.

Yoshua Bengio, Patrice Simard, and Paolo Frasconi. 1994. Learning long-term dependencies with gradient descent is difficult. *IEEE transactions on neural networks*, 5(2):157–166.

Steven Bird, Ewan Klein, and Edward Loper. 2009. *Natural language processing with Python.* " O'Reilly Media, Inc.".

Wendy W Chapman, Dieter Hilert, Sumithra Velupillai, Maria Kvist, Maria Skeppstedt, Brian E Chapman, Michael Conway, Melissa Tharp, Danielle L Mowery, and Louise Deleger. 2013. Extending the NegEx lexicon for multiple languages. *Studies in health technology and informatics*, 192:677.

Danqi Chen and Christopher D Manning. 2014. A fast and accurate dependency parser using neural networks. In *EMNLP*, pages 740–750.

Ronan Collobert and Jason Weston. 2008. A unified architecture for natural language processing: Deep neural networks with multitask learning. In *Proceedings of the 25th international conference on Machine learning*, pages 160–167. ACM.

Ronan Collobert, Jason Weston, Léon Bottou, Michael Karlen, Koray Kavukcuoglu, and Pavel Kuksa. 2011. Natural language processing (almost) from scratch. *Journal of Machine Learning Research*, 12(Aug):2493–2537.

Marie-Catherine De Marneffe, Timothy Dozat, Natalia Silveira, Katri Haverinen, Filip Ginter, Joakim Nivre, and Christopher D Manning. 2014. Universal Stanford dependencies: A cross-linguistic typology. In *LREC*, volume 14, pages 4585–4592.

Sander Dieleman, Jan Schlter, Colin Raffel, Eben Olson, Sren Kaae Snderby, Daniel Nouri, Daniel Maturana,

Martin Thoma, Eric Battenberg, Jack Kelly, Jeffrey De Fauw, Michael Heilman, diogo149, Brian McFee, Hendrik Weideman, takacsg84, peterderivaz, Jon, instagibbs, Dr. Kashif Rasul, CongLiu, Britefury, and Jonas Degrave. 2015. Lasagne: First release., August. Available at http://dx.doi.org/10.5281/zenodo.27878 .

John Duchi, Elad Hazan, and Yoram Singer. 2011. Adaptive subgradient methods for online learning and stochastic optimization. *Journal of Machine Learning Research*, 12(Jul):2121–2159.

Carol Friedman, George Hripcsak, William DuMouchel, Stephen B Johnson, and Paul D Clayton. 1995. Natural language processing in an operational clinical information system. *Natural Language Engineering*, 1(01):83–108.

Alex Graves and Jürgen Schmidhuber. 2005. Framewise phoneme classification with bidirectional LSTM and other neural network architectures. *Neural Networks*, 18(5):602–610.

Alex Graves. 2013. Generating sequences with recurrent neural networks. *arXiv preprint arXiv:1308.0850*.

Saeed Hassanpour and Curtis P Langlotz. 2015. Information extraction from multi-institutional radiology reports. *Artificial intelligence in medicine*.

Sepp Hochreiter and Jürgen Schmidhuber. 1997. Long short-term memory. *Neural computation*, 9(8):1735–1780.

George Hripcsak, John HM Austin, Philip O Alderson, and Carol Friedman. 2002. Use of natural language processing to translate clinical information from a database of 889,921 chest radiographic reports 1. *Radiology*, 224(1):157–163.

David B. Johnson, Ricky K. Taira, Alfonso F. Cardenas, and Denise R. Aberle. 1997. Extracting information from free text radiology reports. *Int. J. Digit. Libr.*, 1(3):297–308, December.

John Lafferty, Andrew McCallum, and Fernando Pereira. 2001. Conditional random fields: Probabilistic models for segmenting and labeling sequence data. In *Proceedings of the eighteenth international conference on machine learning, ICML*, volume 1, pages 282–289.

Guillaume Lample, Miguel Ballesteros, Sandeep Subramanian, Kazuya Kawakami, and Chris Dyer. 2016. Neural architectures for named entity recognition. *arXiv preprint arXiv:1603.01360*.

Curtis P Langlotz. 2006. Radlex: a new method for indexing online educational materials 1. *Radiographics*, 26(6):1595–1597.

Christopher D. Manning, Mihai Surdeanu, John Bauer, Jenny Finkel, Steven J. Bethard, and David McClosky. 2014. The Stanford CoreNLP natural language processing toolkit. In *Association for Computational Linguistics (ACL) System Demonstrations*, pages 55–60.

S McGurk, K Brauer, TV Macfarlane, and KA Duncan. 2014. The effect of voice recognition software on comparative error rates in radiology reports. *The British journal of radiology*.

Saeed Mehrabi, Anand Krishnan, Sunghwan Sohn, Alexandra M Roch, Heidi Schmidt, Joe Kesterson, Chris Beesley, Paul Dexter, C Max Schmidt, Hongfang Liu, et al. 2015. DEEPEN: A negation detection system for clinical text incorporating dependency relation into NegEx. *Journal of biomedical informatics*, 54:213–219.

Grégoire Mesnil, Xiaodong He, Li Deng, and Yoshua Bengio. 2013. Investigation of recurrent-neural-network architectures and learning methods for spoken language understanding. In *INTERSPEECH*, pages 3771–3775.

Tomas Mikolov and Geoffrey Zweig. 2012. Context dependent recurrent neural network language model. In *SLT*, pages 234–239.

Tomas Mikolov, Ilya Sutskever, Kai Chen, Greg S Corrado, and Jeff Dean. 2013. Distributed representations of words and phrases and their compositionality. In *Advances in neural information processing systems*, pages 3111–3119.

England NHS. 2016. Diagnostic Imaging Dataset Statistical Release. Available at https://www.england.nhs.uk/statistics/statistical-work-areas/diagnostic-imaging-dataset/diagnostic-imaging-dataset-2015-16-data/.

United States National Library of Medicine NLM. 2016a. Medical Subject Headings. Available at https://www.nlm.nih.gov/mesh/.

United States National Library of Medicine NLM. 2016b. Unified Medical Language System. Available at https://uts.nlm.nih.gov/home.html.

Philip V Ogren, Guergana K Savova, Christopher G Chute, et al. 2007. Constructing evaluation corpora for automated clinical named entity recognition. In *Medinfo 2007: Proceedings of the 12th World Congress on Health (Medical) Informatics; Building Sustainable Health Systems*, page 2325. IOS Press.

Naoaki Okazaki and Jun'ichi Tsujii. 2010. Simple and efficient algorithm for approximate dictionary matching. In *Proceedings of the 23rd International Conference on Computational Linguistics*, pages 851–859. Association for Computational Linguistics.

Jeffrey Pennington, Richard Socher, and Christopher D Manning. 2014. Glove: Global vectors for word representation. In *EMNLP*, volume 14, pages 1532–1543.

Bruce Reiner and Eliot Siegel. 2006. Radiology reporting: returning to our image-centric roots. *American Journal of Roentgenology*, 187(5):1151–1155.

26

David E Rumelhart, Geoffrey E Hinton, and Ronald J Williams. 1988. Learning representations by back-propagating errors. *Cognitive modeling*, 5(3):1.

Guergana K Savova, James J Masanz, Philip V Ogren, Jiaping Zheng, Sunghwan Sohn, Karin C Kipper-Schuler, and Christopher G Chute. 2010. Mayo clinical text analysis and knowledge extraction system (cTAKES): architecture, component evaluation and applications. *Journal of the American Medical Informatics Association*, 17(5):507–513.

Hoo-Chang Shin, Le Lu, Lauren Kim, Ari Seff, Jianhua Yao, and Ronald M Summers. 2015. Interleaved text/image deep mining on a very large-scale radiology database. In *Proceedings of the IEEE Conference on Computer Vision and Pattern Recognition*, pages 1090–1099.

Chaitanya Shivade, Marie-Catherine de Marneffe, Eric Fosler-Lussier, and Albert M Lai. 2015. Extending NegEx with kernel methods for negation detection in clinical text. *ExProM 2015*, page 41.

Allan F Simpao, Luis M Ahumada, Jorge A Gálvez, and Mohamed A Rehman. 2014. A review of analytics and clinical informatics in health care. *Journal of medical systems*, 38(4):1–7.

Jeffrey L Sobel, Marjorie L Pearson, Keith Gross, Katherine A Desmond, Ellen R Harrison, Lisa V Rubenstein, William H Rogers, and Katherine L Kahn. 1996. Information content and clarity of radiologists' reports for chest radiography. *Academic radiology*, 3(9):709–717.

Pontus Stenetorp, Sampo Pyysalo, Goran Topić, Tomoko Ohta, Sophia Ananiadou, and Jun'ichi Tsujii. 2012. BRAT: a web-based tool for NLP-assisted text annotation. In *Proceedings of the Demonstrations at the 13th Conference of the European Chapter of the Association for Computational Linguistics*, pages 102–107. Association for Computational Linguistics.

Theano Development Team. 2016. Theano: A Python framework for fast computation of mathematical expressions. *arXiv e-prints*, abs/1605.02688, May.

Paul J Werbos. 1990. Backpropagation through time: what it does and how to do it. *Proceedings of the IEEE*, 78(10):1550–1560.

Stephen Wu, Timothy Miller, James Masanz, Matt Coarr, Scott Halgrim, David Carrell, and Cheryl Clark. 2014. Negations not solved: generalizability versus optimizability in clinical natural language processing. *PloS one*, 9(11):e112774.

Mo Yu and Mark Dredze. 2014. Improving lexical embeddings with semantic knowledge. In *ACL (2)*, pages 545–550.

Data Resource Acquisition from People at Various Stages of Cognitive Decline – Design and Exploration Considerations

Dimitrios Kokkinakis
Department of Swedish
University of Gothenburg
Sweden
dimitrios.kokkinakis@gu.se

Kristina Lundholm Fors
Department of Swedish
University of Gothenburg
Sweden
kristina.lundholm@gu.se

Arto Nordund
Department of Psychiatry
and Neurochemistry
University of Gothenburg
Sweden
arto.nordlund@gu.se

Abstract

In this paper we are introducing work in progress towards the development of an infrastructure (i.e., design, methodology, creation and description) of linguistic and extra-linguistic data samples acquired from people diagnosed with subjective or mild cognitive impairment and healthy, age-matched controls. The data we are currently collecting consists of various types of modalities; i.e. audio-recorded spoken language samples; transcripts of the audio recordings (text) and eye tracking measurements. The integration of the extra-linguistic information with the linguistic phenotypes and measurements elicited from audio and text, will be used to extract, evaluate and model features to be used in machine learning experiments. In these experiments, classification models that will be trained, that will be able to learn from the whole or a subset of the data to make predictions on new data in order to test how well a differentiation between the aforementioned groups can be made. Features will be also correlated with measured outcomes from e.g. language-related scores, such as word fluency, in order to investigate whether there are relationships between various variables.

1 Introduction

Current state-of-the-art diagnostic measures for Alzheimer's Disease (AD) are invasive, expensive, and time-consuming. There is a consensus on the need for the identification of the disease in its earliest manifestations by applying non-invasive and/or cost-effective methods that could aid the identification of subjects in the preclinical or early clinical stages. Efficient tools that could be applied in routine dementia screening in primary care settings for identifying subjects who could be appropriate for further cognitive evaluation and dementia diagnostics[1], could provide the specialist centres the opportunity to engage in more demanding, advanced investigations, care and treatment. New paths of research traced to acquire further knowledge about AD and its subtypes as well as tools based on the exploration of several complementary modalities and parameters, such as speech analysis and/or eye testing (cf. Laske et al., 2014, for a review) could be examined and incorporated into established neuropsychological, memory and cognitive tests in order to investigate fairly unexplored features that may be used as potential biomarkers for AD. This paper describes some efforts underway to acquire, assess, analyze and evaluate linguistic and extra-linguistic data from people with subjective (SCI) and mild cognitive impairment (MCI) and healthy, age-matched controls.

The SCI, the MCI, and the Alzheimer's disease (AD) are on a spectrum of disease progression. Subjective cognitive impairment (SCI) is a common diagnosis in elderly people, sometimes suggested to be associated with e.g. depression, stress or anxiety,

---

[1] Currently in e.g. Sweden (Nordberg et al., 2014) only 30% of all Alzheimer's disease receive a complete memory investigation, diagnosis and symptomatic drugs, the rest of the cases are assigned the codes UNS, "unspecified dementia".

Proceedings of the Seventh International Workshop on Health Text Mining and Information Analysis (LOUHI), pages 28–36,
Austin, TX, November 5, 2016. ©2016 Association for Computational Linguistics

but also a risk factor for dementia (Jessen et al., 2010). On the other hand, mild cognitive impairment (MCI) is a prodromal state of dementia (Ritchie & Touchon, 2010), in which someone has minor problems with cognition (e.g., problems with memory or thinking) but these are not severe enough to warrant a diagnosis of dementia or interfere significantly with daily life, but still the difficulties are worse than would normally be expected for a healthy person of their age.

The rest of this paper is organized as follows: Section 2 presents related work from the Computational Linguistic/Natural Language Processing (CL/NLP) field in the domain of dementia, with focus on classification and prediction methodologies using mainly spoken language (including transcribed data) as well as eye tracking measures. Sections 3 and 4 provide a description of the protocol used in the project. Section 3 briefly discusses the Gothenburg MCI-study (from which the current project is recruiting its subjects) and also the ethical issues related to the project, while Section 4 presents the material and design of the various experimental tasks and the procedure for data collection. Post-processing of acquired data is also discussed in Section 4 while Section 5 provides a brief outline of the features we plan to extract from the data and the algorithms to use for classification and statistical analysis. Finally, in Section 6, the conclusions and future work are presented.

2   CL/NLP and the area of Dementia

A prerequisite for identifying dementia in its earliest stages is a reliable cognitive examination (Nordlund et al., 2010). Particularly for clinicians, language plays an important role in diagnosis and investigations include inquiries about the use of language in various forms. New findings aim to provide a comprehensive picture of cognitive status and promising results have recently thrown more light on the importance of language and language (dis)abilities as an essential factor that can have a strong impact on specific measurable characteristics that can be extracted by automatic linguistic analysis of speech and text (Ferguson et al., 2013; Szatloczki et al., 2015). The work by Snowdon et al. (2000), "The Nun Study", was one of the earliest studies which showed a strong correlation between low linguistic ability early in life and cognitive impairment in later life by analyzing autobiographies of American nuns and could predict who could develop Alzheimer's Dementia by studying the degradation of the idea density (that is, the average number of ideas expressed in 10 words; Chand et al., 2010) and syntactic complexity in the nuns' autobiographical writings.

Since then, the body of research and interest in CL/NLP research in the area of processing data from subjects with mental, cognitive, neuropsychiatric, or neurodegenerative impairments has grown rapidly[2]. Automatic spoken language analysis and eye movement measurements are two of the newer complementary diagnostic tool with great potential for dementia diagnostics (Laske et al., 2014). Furthermore, the identification of important linguistic and extra-linguistic features such as lexical and syntactic complexity, are becoming an established way to train and test machine learning classifiers that can be used to differentiate between subjects with various forms of dementia and healthy controls or between different types of dementia subjects (Lagun et al., 2011; Roark et al., 2011; Olubolu Orimaye et al., 2014; Rentoumi et al., 2014).

Although language is not the only diagnostic factor for cognitive impairment, several recent studies (Yancheva et al., 2015) have demonstrated that automatic linguistic analysis, primarily of speech samples, produced by people with mild or moderate cognitive impairment compared to healthy individuals can identify objective evidence and measurable (progressive) language disorders. Garrard & Elrevåg (2014) comment that computer-assisted analysis of large language datasets could contribute to the understanding of brain disorders. Although, none of the studies presented in the special issue of Cortex vol. 55 moved "beyond the representation of language as text" and therefore finding reliable ways of incorporating features, such as prosody and

---

[2] See e.g. the three "Computational Linguistics and Clinical Psychology" workshops (<http://clpsych.org/>); the LREC workshop on "Resources and Processing of linguistic and extra-linguistic data from people with various forms of cognitive/psychiatric impairments", RaPID (<https://spraak-banken.gu.se/eng/rapid-2016>); various papers in the workshop series on "Speech and Language Processing for Assistive Technologies", SLPAT (<http://www.slpat.org/>) and the seven Louhi: Workshops on Health Text Mining and Information Analysis (<https://louhi.limsi.fr/2016/>).

emotional connotation, into data representation remains a future challenge, the editors acknowledged that current research indicates that "the challenges of applying computational linguistics to the cognitive neuroscience field, as well as the power of these techniques to frame questions of theoretical interest and define clinical groups are of practical importance". Nevertheless, studies have shown that a steady change in the linguistic nature and the degree of symptoms in speech and writing are early and could be identified by using language technology analysis (Mortimer et al., 2005; Le et al., 2011). New findings also show a great potential to increase our understanding of dementia and its impact on linguistic degradation such as loss of vocabulary, syntactic simplification, poor speech content and semantic generalization. Analysis of eye movement is also a relevant research technology to apply, and text reading by people with and without mild cognitive impairment may give a clear ruling on how reading strategies differ between these groups, an area that has so far not been researched to any significant extent in this particular domain (Fernández et al., 2013, 2014; Molitor et al., 2015). With the help of eye-tracking technology the eye movements of participants are recorded while suitable stimuli is presented (e.g., a short text; cf. section 4.3).

3 The Gothenburg MCI-study and Related Ethical Issues

The ongoing Gothenburg mild cognitive impairment study (Nordlund et al., 2005; Wallin et al., 2016) is an attempt to conduct longitudinal in-depth phenotyping of patients with different forms and degrees of cognitive impairment using neuropsychological, neuroimaging, and neurochemical tools. The study is clinically based and aims at identifying neurodegenerative, vascular and stress related disorders prior to the development of dementia. All patients in the study undergo baseline investigations, such as neurological examination, psychiatric evaluation, cognitive screening (e.g., memory and visuospatial disturbance, poverty of language and apraxia), magnetic resonance imaging of the brain and cerebrospinal fluid collection. At biannual follow-ups, most of these investigations are repeated.

The overall Gothenburg MCI-study is approved by the local ethical committee review board (reference number: L091–99, 1999; T479-11, 2011); while the currently described study by the local ethical committee decision 206-16, 2016). The project aims at gathering a rather homogeneous group of participants with respect to age and education level (50 with SCI/MCI and 50 controls). Written informed consent is obtained from all participants in the study while the exclusion and inclusion criteria are specified according to the following:

Inclusion criteria

- Age 50-79 years
- Swedish as a first language and not speaking languages other than Swedish before the age of 5
- Comparable education length of the participants
- No apparent organic cause of symptoms
- Research subjects have read information about the research project[3] and approved voice recording and eye movement measurements

Exclusion criteria

- Participants have dyslexia or other reading difficulties
- Participants have deep depression
- Participants have an ongoing abuse of any kind
- Participants suffer from serious psychiatric or neurological diseases such as Parkinson's, Amyotrophic lateral sclerosis or have/had a brain tumor
- Participants do not understand the question or the context in the selection process
- Participants have poor vision (that cannot be corrected by glasses or lenses), cataract, nystagmus, or cannot see and read on the computer screen

---

[3] According to the instructions provided by the Swedish ethical review board <http://www.epn.se/media/1210/information_for_research_participants.pdf>.

- Participants decline participation during telephone call or later at the recording site
- Participants decline signing the paper of informed consent
- Recordings or eye movement measurements are technically unusable.

4 Material and Design of Experiments

The purpose of the acquisition of the data (audio recordings, transcriptions[4] and eye tracking measurements) is to facilitate feature extraction in machine learning experiments (see Section 5).

4.1 Audio Recordings

For the acquisition of the audio signal we use the Cookie-theft picture (see Figure 1) from the *Boston Diagnostic Aphasia Examination* (BDAE; Goodglass & Kaplan, 1983) which is often used to elicit speech from people with various mental and cognitive impairments. During the presentation of the Cookie Theft stimuli (which illustrates an event taking place in a kitchen) the subjects are asked to tell a story about the picture and describe everything that can be observed while the story is recorded. For the task the original label of the cookie jar is translated and substituted from the English "COOKIE JAR" to the Swedish label "KAKBURK". The picture is considered an "ecologically valid approximation" to spontaneous discourse (Giles et al., 1996).

Figure 1: The Cookie Theft picture

We chose to use the Cookie Theft picture[5] since it provides a standardized test that has been used in various studies in the past, therefore comparisons can be made based on previous results, e.g. with research on the DementiaBank database or other collections (MacWhinney, 2007; Williams et al., 2010; Fraser & Hirst, 2016). Moreover, in order to allow the construction of a comprehensive speech profile for each research participant, the speech task also includes reading aloud a short text from the *International Reading Speed Texts* collection (IReST; Trauzettel-Klosinski et al., 2012) presented on a computer screen. As a matter of fact, two texts are used from this collection, in connection to the eye tracking experiment (see next section), but only one of those texts is read aloud and thus combined with eye-tracking recording; cf. Meilán et al., 2012 and 2014 for similar "reading out" text passage experiments. IReST is a multilingual standardized text collection used to assess reading performance, for multiple equivalent texts for repeated measurements. Specifically in our project we use the Swedish IReST translations, namely texts "one" and "seven" (Öqvist Seimyr, 2010). For the audio capture of both we use a H2n Handy recorder[6] while the audio files are saved and stored as uncompressed audio in .wav 44.1 kHz/16 bit format. Recordings are carried out in an isolated environment in order to avoid noise.

4.2 Verbatim Transcriptions

The textual part of the infrastructure consists of manually produced transcriptions of the two audio recordings previously described. The digitized speech waveform will be semi-automatically aligned with the transcribed text. During transcription, special attention will also be paid to non-speech acoustic events including speech dysfluencies consisting of filled pauses a.k.a. hesitation ("um"), false-starts, repetitions as well as other features such as laughing. A very basic transcription manual is also produced which will help the human transcribers accomplish a homogeneous transcription. Furthermore, for the transcription the PRAAT application (Boersma & Weenink, 2013) is used.

---

[4] Since some of the features to be extracted (e.g. part-of-speech and syntactic labels from the transcriptions) are language-dependent it requires the use of a language-specific infrastructure (in our case Swedish), for that reason we plan to use available resources; cf. Ahlberg et al. (2013); possible modifications to the transcribed language are also envisaged.

[5] The Cookie theft picture, but with written descriptions of it, has also been used in a few studies with Swedish subjects (cf. Tyche 2001; Cromnow & Landberg, 2009; Landfeldt & Söderbäck, 2009). In all these studies the analysis was based on narrative writing of the Cookie theft picture.

[6] From ZOOM Corp. <https://www.zoom-na.com/sv>.

4.3 Eye-Tracking

The investigation of eye movement functions in SCI/MCI, and any differences or changes in eye movements that could be potentially detected for those patients is of great importance to clinical AD research. However, until now, eye tracking has not been used to investigate reading for MCI-persons in a much larger scale, possibly due to the number of procedural difficulties related to this kind of research. On the other hand, the technology has been applied in a growing body of various experiments related to other impairments such as autism (Yaneva et al., 2016; Au-Yeung et al., 2015) and dyslexia (Rello & Ballesteros, 2015). For the experiments we use EyeLink[7] 1000 Desktop Mount with monocular eye tracking with head stabilization and a real-time sample access of 1000Hz. Head stabilization provides an increased eye tracking range performance. The participants were seated in front of the monitor at a distance of 60-70 cm. While reading, the eye movements of the participants are recorded with the eye-tracking device while interest areas around each word in the text are defined by taking advantage of the fact that there are spaces between each word in the text. The eye-tracking measurements are used for the detection and calculation of fixations, saccades and backtracks. Fixation analyses is conducted within predefined Areas of Interest (AOI); in our case each word is an AOI.

4.4 Comparison over a Two-Year Span

The previously outlined experiments/recordings will be repeated two years after the first recording taking place during the second half of 2016. This way we want to analyze whether there are any differences and at which level and magnitude between the two audio and eye-tracking recordings. Namely, compare and examine whether there any observable, greater, differences/decline on some features and which these could be. We are aware that more longitudinal data samples over a longer time period would be desirable but at this stage only a single repetition is practically feasible to perform. In other, longitudinal experiments, e.g. in investigating the nature and progression of the spontaneous writing, patterns of impairment were observed in patients with Alzheimer's disease over a 12-month period,

these were dominated by semantic errors (Forbes-McKay et al., 2014).

5 Envisaged Analysis and Features

The envisaged analysis and exploration intends to extract, evaluate and combine a number of features from the three modalities selected to be investigated. These are speech-related features, text/transcription-related features and eye tracking-related features.

5.1 Speech-related Features

A large number of acoustic features have been proposed in the literature which pinpoints of the importance of distinguishing between vocal changes that occur with normal aging and those that are associated with MCI (and AD). We expect that our spoken samples will show different features depending on whether they are produced spontaneously (when talking about the Cookie theft picture) or they consist of read aloud speech. Reliable and robust acoustic features that might differentiate spoken language in SCI/MCI and healthy controls remains an ongoing challenge but the technology develops rapidly. Roark et al. (2011) used 21 features in supervised machine learning experiments (using Support Vector Machines) from 37 MCI subjects and equally many controls (37/37). Features from both the audio and the transcripts included: pause frequency, filled pauses, total pause duration and linguistic variables such as Frazier and Yngve scores and idea density, while best accuracy with various feature configurations were 86.1% for the area under the ROC curve. Pause frequency has been identified as a feature differentiating spontaneous speech in patients with AD from control groups (Gayraud et al., 2011), and may also be used to distinguish between mild, moderate and severe AD (Hoffman et al., 2010). Meilán et al. (2014) used AD subjects and spoken data (read loud and clear sentences on a screen). They used acoustic measures such as pitch, volume and spectral noise measures. Their method was based on linear discriminant analysis and their results could characterize people with AD with an accuracy of 84.8%. Yancheva et al. (2015) used spoken and transcriptions features provided from the DementiaBank (Cookie theft descriptions) using 393 speech samples (165/90).

---

[7] From SR Research Ltd. <http://www.sr-research.com/>.

They extracted and investigated 477 different features both lexicosyntactic ones (such as syntactic complexity; word types, quality and frequency) and acoustic ones (such as Melfrequency cepstral coefficients – MFCC, including mean, variance, skewness, and kurtosis; pauses and fillers; pitch and formants and aperiodicity measures) and semantic ones (such as concept mention) in order to predict Mini Mental State Examination (MMSE[8]) scores with a mean absolute error of 3.83 while with individuals with more longitudinal samples the mean absolute error was improved to 2.91, which suggested that the longitudinal data collection plays an important role. König et al. (2015) looked also at MCI and AD subjects (23/26) and examined vocal features (silence, voice, periodic and aperiodic segment length; mean of durations) using Support vector machine (SVM). Their classification accuracy of automatic audio analyses was 79% between healthy controls and those with MCI and 87% between healthy controls and those with AD; and between those with MCI and those with AD, 80%. Tóth et al. (2015) used also SVM and achieved 85.3% F-score (32 MCI subjects and 19 controls) by starting with eight acoustic features extracted by applying automatic speech recognition (such as speech tempo i.e. phones per second) and extending them to 83. Finally, Fraser et al. (2016) also looked at the DementiaBank and using 240 samples of AD and 233 from healthy controls, extracted 370 features, such as linguistic variables from transcripts (e.g., part-of-speech frequencies; syntactic complexity and grammatical constituents), psycholinguistic measures (e.g., vocabulary richness) and acoustic variables from the audio files (e.g., MFCC). Using logistic regression, Fraser et al. could obtain a classification accuracy of 81% in distinguishing individuals with AD from those without based on short samples of their language on the Cookie Theft picture description task.

5.2 Text/Transcription-related Features

Many of the previous studies combine both acoustic features and features from the transcriptions; cf. the supplementary material in Fraser et al. (2016). Some of the most common features and measures from transcribed text follow the lexicon-syntax-semantics continuum. These measures include (i) *lexical distribution measures* (such as type-token ratio, mean word length, long word counts, hapax legomena, hapax dislegomena, automated readability index and Coleman-Liau Index; also lexical and non-lexical fillers or disfluency markers, i.e. "um", "uh", "eh") and out-of-vocabulary rate (Pakhomov et al., 2010). (ii) *syntactic complexity markers* (such as frequency of occurrence of the most frequent words and deictic markers; [context free] production rules, i.e. the number of times a production rule is used divided by the total number of productions; dependency distance, i.e. the length of a dependency link between a dependent token and its head, calculated as the difference between their positions in a sentence; parse tree height, i.e. is the mean number of nodes from the root to the most distant leaf; depth of a syntactic tree, i.e. the proportion of subordinate and coordinate phrases to the total number of phrases and ratio of subordinate to coordinate phrases; noun phrase average length and noun phrase density, i.e. the number of noun phrases per sentence or clause; words per clause); and (iii) *semantic measures* (such as the idea or propositional density, i.e. the operationalization of conciseness – the average number of ideas expressed per words used; the number of expressed propositions divided by the number of words; a measure of the extent to which the speaker is making assertions, or asking questions, rather than just referring to entities etc.).

5.3 Eye Tracking-related Features

Eye tracking data has been used in machine learning methods in the near past that take advantage of eye dynamics biomarkers (Lagun et al; 2011) with good indication that they can aid the automatic detection of cognitive impairment (i.e., distinguish healthy controls from MCI-patients). Several studies provide evidence and suggest that eye movements can be used to detect memory impairment and serve as a possible biomarker for MCI and, in turn, AD (Fernández et al., 2013). Basic features we intend to investigate in this study are *fixations* (that is the state the eye remains still over a period of time); *saccades* (that is the rapid motion of the eye from one fixation to another) and *backtracks* (that is the relationship between two subsequent saccades where the second goes in the opposite direction than the first); for a

---

[8] MMSE is a brief screening test that quantitatively estimates the severity and progression of cognitive impairment and also cognitive changes over time (Tombaugh & McIntyre, 1992).

thorough description of possible eye-tracking related features cf. Holmqvist et al., 2015:262. Saccades are of particular interest because they are much related to attention and thus, they are likely to be disturbed by cognitive impairments associated with neurodegenerative disorders (Anderson et al., 2013). Note that there are many assumptions behind the use of eye tracking technology for experiments designed for people with MCI. For instance, the longer the eye gaze fixation is on a certain word, the more difficult the word is for cognitive processing, therefore the durations of gaze fixations could be used as a proxy for measuring cognitive load (Just & Carpenter, 1980). Molitor et al. (2015) provide a recent review on the growing body of literature that investigates changes in eye movements as a result of AD and the alterations to oculomotor function and viewing behavior.

5.4 Correlation Analysis

We intend to further perform correlation analysis with the features previously outlined and the results from various measures from language-related tests performed in the Gothenburg MCI-study, tests which are applied for assessing possible dementia. Typically, clinicians use tests such as Mini-Mental State Examination (MMSE), linguistic memory tests and language tests. Such tests include the token test, subtest V, which is a test of syntax comprehension; the Boston naming test and the word fluency FAS test (the number of words initiated by the letters F, A, and S). This investigation intends to identify whether there are variables/features (highly) correlated with i.e. the MCI class, yet uncorrelated with each other i.e. the healthy controls or SCI.

6 Conclusions and Future Work

In this paper we have introduced work in progress towards the design and infrastructural development of reliable multi-modal data resources and a set of measures (features) to be used both for experimentation with feature engineering and evaluation of classification algorithms to be used for differentiating between SCI/MCI and healthy adults, and also as benchmark data for future research in the area. Evaluation practices are a crucial step towards the development of resources and useful for enhancing progress in the field, therefore we intend to evaluate both the relevance of features, compare standard algorithms such as Support vector machine and Bayesian classifiers and perform correlation analysis with the results of established neuropsychological, memory and cognitive tests. We also intend to repeat the experiments after two years in order to assess possible changes at each level of analysis. We believe that combining data from three modalities could be useful, but at this point we do not provide any clinical evidence underlying these assumptions since the analysis and experimentation studies are planned for year 2 of the project (2017). Therefore, at this stage, the paper only provides a snapshot of the current stage of the work.

Acknowledgments

This work has received support from *Riksbankens Jubileumsfond* - The Swedish Foundation for Humanities and Social Sciences, through the grant agreement no: NHS 14-1761:1.

References

Malin Ahlberg et al. 2013. *Korp and Karp – a bestiary of language resources: the research infrastructure of Språkbanken*. Proceedings of the 19th Nordic Conference of Computational Linguistics (NODALIDA). Linköping Electronic Conference Proceedings #85.

Tim J. Anderson and Michael R. MacAskill. 2013. Eye movements in patients with neurodegenerative disorders. *Nat Rev Neurology* 9: 74-85. doi:10.1038/nrneurol.2012.273.

Sheena K. Au-Yeung, Johanna Kaakinen, Simon Liversedge and Valerie Benson. 2015. Processing of Written Irony in Autism Spectrum Disorder: An Eye-Movement Study. *Autism Res.* 8(6):749-60. doi: 10.1002/aur.1490.

Paul Boersma and David Weenink. 2013. *Praat: doing phonetics by computer* [Computer program]. Version 6.0.19, retrieved in Aug. 2016 from <http://www.praat.org/>.

Vineeta Chand, Kathleen Baynes, Lisa M. Bonnici and Sarah Tomaszewski Farias. 2012. A Rubric for Extracting Idea Density from Oral Language Samples Analysis of Idea Density (AID): A Manual. In *Curr Protoc Neurosci.* CHAPTER: Unit10.5. doi:10.1002/0471142301.ns1005s58.

Karolina Cromnow and Tove Landberg. 2009. *Skriftliga beskrivningar av bilden "Kakstölden". Insamling av referensvärden från friska försökspersoner*. Master's Thesis. Department of Clinical Sciences, division of Speech and Language Pathology. Karolinska Institute, Sweden. (In Swedish).

Alison Ferguson, Elizabeth Spencer, Hugh Craig and Kim Colyvas. 2014. Propositional Idea Density in women's written language over the lifespan: Computerized analysis. *Cortex* 55. 107-121. dx.doi.org/10.1016/j.cortex.2013.05.012

Gerardo Fernández, Pablo Mandolesi, Nora P. Rotstein, Oscar Colombo, Osvaldo Agamennoni and Luis E. Politi. 2013. Eye Movement Alterations During Reading in Patients With Early Alzheimer Disease. *Investigative Ophthalmology & Visual Science*. Vol.54, 8345-8352. doi:10.1167/iovs.13-12877.

Gerardo Fernández, Jochen Laubrock, Pablo Mandolesi, Oscar Colombo and Osvaldo Agamennoni. 2014. Registering eye movements during reading in Alzheimer's disease: difficulties in predicting upcoming words. *J of Clin & Experimental Neuropsychol.* 36(3):302–16.

Katrina Forbes-McKay, Mike Shanks and Annalena Venneria. 2014. Charting the decline in spontaneous writing in Alzheimer's disease: a longitudinal study. *Acta Neuropsychiatrica*. Vol. 26:04, pp 246-252 doi: http://dx.doi.org/10.1017/neu.2014.2.

Kathleen C. Fraser and Graeme Hirst. 2016. *Detecting semantic changes in Alzheimer's disease with vector space models*. LREC Workshop: RaPID (Resources and ProcessIng of linguistic and extra-linguistic Data from people with various forms of cognitive/psychiatric impairments). Pp. 1-8. Portorož Slovenia.

Kathleen C. Fraser, Jed A. Meltzer and Frank Rudzicz. 2016. Linguistic Features Identify Alzheimer's Disease in Narrative Speech. *J of Alzheimer's Disease* 49 (2016) 407–422. IOS Press. doi:10.3233/JAD-150520

Peter Garrard and Brita Elvevåg. 2014. Special issue: Language, computers and cognitive neuroscience. *Cortex* 55; 1-4. DOI: 10.1016/j.cortex.2014.02.012.

Frederique Gayraud, Hye-Ran Lee and Melissa Barkat-Defradas. 2011. Syntactic and lexical context of pauses and hesitations in the discourse of Alzheimer patients and healthy elderly subjects. *Clinical Linguistics & Phonetics*, *25*(3), 198–209. doi: 10.3109/02699206.2010.521612.

Elaine Gilles, Karalyn Patterson and John R. Hodges. 1996. Performance on the Boston Cookie Theft picture description task in patients with early dementia of the Alzheimer's type: missing information. *Aphasiology*. Vol. 10:4. Pp. 395-408.

Harold Goodglass and Edith Kaplan. 1983. *The Assessment of Aphasia and Related Disorders*. 2nd edition. Lea & Febiger, Philadelphia, PA, USA.

Ildikó Hoffmann, Dezso Nemeth, Cristina D Dye, Magdolna Pákáski, Tamás Irinyi and János Kálmán. 2010. Temporal parameters of spontaneous speech in Alzheimer's disease. *J of Speech-Language Pathology*, *12*(1), 29–34. doi: 10.3109/17549500903137256.

Kenneth Holmqvist, Marcus Nyström, Richard Andersson, Richard Dewhurst, Halszka Jarodzka, and Joost van de Weijer. 2015. *Eye Tracking - A comprehensive guide to methods and measures*. OUP.

Frank Jessen et al. 2010. Prediction of dementia by subjective memory impairment: effects of severity and temporal association with cognitive impairment. *Arch. Gen. Psychiatry*, 67(4):414–422. doi:10.1001/archgenpsychiatry.2010.30.

Marcel A. Just and Patricia A. Carpenter. 1980. A theory of reading: from eye fixations to comprehension. *Psychological review*, 87(4):329-354.

Alexandra König, Aharon Satt, Alexander Sorin, Ron Hoory, Orith Toledo-Ronen, Alexandre Derreumaux, Valeria Manera, Frans Verhey, Pauline Aalten, Phillipe H. Robert and Renaud David. 2015. Automatic speech analysis for the assessment of patients with predementia and Alzheimer's disease. *Alzheimer's & Dementia: Diagnosis, Assessment & Disease Monitoring*. 1:112–124. Elsevier.

Dmitry Lagun, Cecelia Manzanares, Stuart M. Zola, Elizabeth A. Buffalo and Eugene Agichtein. 2011. Detecting cognitive impairment by eye movement analysis using automatic classification algorithms. *J Neurosci Methods*. 201(1): 196–203.

Erik Landfeldt and Emma Söderbäck. 2009. *Predicerar skriftliga bildbeskrivningar demens? - En retrospektiv studie*. Master's Thesis. Department of Neuroscience, Division of Speech and Language Pathology. Uppsala university. Sweden. (In Swedish).

Christoph Laske et al. 2014. Innovative diagnostic tools early detection of Alzheimer's disease. *Alzheimer's & Dementia*. 1-18. DOI: 10.1016/j.jalz.2014.06.004.

Xuan Le, Ian Lancashire, Graeme Hirst, and Regina Jokel. 2011. Longitudinal Detection of Dementia through Lexical and Syntactic Changes in Writing: A Case Study of Three British Novelists. *J of Literary and Linguistic Computing*. 26 (4): 435-461

Brian MacWhinney. 2007. *The Talkbank Project*. In Creating and Digitizing Language Corpora. J.C. Beal (eds). Pp. 163–180. Springer.

Juan JG. Meilán, Francisco Martínez-Sánchez, Juan Carro, José A. Sánchez and Enrique Pérez. 2012. Acoustic Markers Associated with Impairment in Language Processing in Alzheimer's Disease. Spanish *Journal of Psychology*. Vol. 15:2, 487-494 dx.doi.org/10.5209/rev_SJOP.2012.v15.n2.38859

Juan JG. Meilán, Francisco Martínez-Sánchez, Juan Carro, Dolores E. López, Lymarie Millian-Morell and José M. Arana. 2014. Speech in Alzheimer's Disease: Can Temporal and Acoustic Parameters Discriminate Dementia? *Dement Geriatr Cogn Disord* 2014;37:327–334. doi: 10.1159/000356726

Robert J. Molitor, Philip C. Ko and Brandon A. Ally. 2015. Eye Movements in Alzheimer's Disease. *J of Alzheimer's Disease* 44, 1–12. doi:10.3233/JAD-141173. IOS Press.

James A. Mortimer, Amy R. Borenstein, Karen M. Gosche and David A. Snowdon. 2005. Very Early Detection of Alzheimer Neuropathology and the Role of Brain Reserve in Modifying Its Clinical Expression. *J Geriatr Psychiatry Neurol.* 18(4): 218–223.

Agneta Nordberg et al. 2014. *Alzheimerdrabbade ges inte samma rätt som andra sjuka.* Publiced 26/7/2014 in Dagens Nyheter Debatt: http://www.dn.se/debatt/alzheimerdrabbade-ges-inte-samma-ratt-somandrasjuka/ (In Swedish)

Arto Nordlund, Sindre Rolstad, Per Hellström, Magnus Sjögren, Stefan Hansen and Anders Wallin. 2005. The Goteborg MCI study: mild cognitive impairment is a heterogeneous condition. *J Neurol Neurosurg Psych.* 76(11): 1485–1490. doi: 10.1136/jnnp.2004.050385.

Arto Nordlund et al. 2010. Cognitive profiles of incipient dementia in the Goteborg MCI study. *Dement Geriatr Cogn Disord.* 30(5):403-10. doi: 10.1159/000321352.

Sylvester Olubolu Orimaye, Jojo Sze-Meng Wong and Kren J. Golden. 2014. *Learning Predictive Linguistic Features for Alzheimer's Disease and related Dementias using Verbal Utterances.* Workshop on Computational Ling. & Clinical Psychology: From Linguistic Signal to Clinical Reality. 78–87. Maryland, USA.

Serguei VS Pakhomov, Glenn E. Smith, Susan Marino, Angela Birnbaum, Neill Graff-Radford, Richard Caselli, Bradley Boeve and David Knopman. 2010. A computerized technique to assess language use patterns in patients with frontotemporal dementia. *J Neuroling.* 23(2):127–144.doi:10.1016/j.jneuroling.2009.12.001.

Luz Rello and Miguel Ballesteros. 2015. *Detecting Readers with Dyslexia Using Machine Learning with Eye Tracking Measures.* Proceedings of the 12th Web for All Conference W4A. Florence, Italy. doi: 10.1145/2745555.2746644.

Vassiliki Rentoumi, Ladan Raoufian, Samrah Ahmed and Peter Garrard. 2014. Features and Machine Learning Classification of Connected Speech Samples from Patients with Autopsy Proven Alzheimer's Disease with and without Additional Vascular Pathology. *J of Alzheimer's Disease* 42. IOS Press. S3–S17. doi: 10.3233/JAD-140555.

Karen Ritchie and Jacques Touchon. 2010. Mild cognitive impairment: conceptual basis and current nosological status. *The Lancet.* Vol. 355:9199. Pp. 225–228. doi:10.1016/S0140-6736(99)06155-3.

Brian Roark, Margaret Mitchell, John-Paul Hosom, Kristy Hollingshead, and Jeffrey Kaye. 2011. *Spoken Language Derived Measures for Detecting Mild Cognitive Impairment.* IEEE Trans Audio Speech Lang Processing. 19(7): 2081–2090

David A. Snowdon, Lydia Greiner and William R. Markesbery. 2000. Linguistic ability in early life and the neuropathology of Alzheimer's disease and cerebrovascular disease. Findings from the Nun Study. *Annals of the New York Acad of Sciences.* 903:34-8.

Greta Szatloczki, Ildiko Hoffmann, Veronika Vincze, Janos Kalman and Magdolna Pakaski. 2015. Speaking in Alzheimer's disease, is that an early sign? Importance of changes in language abilities in Alzheimer's disease. *Frontiers in Aging Neuroscience.* Vol 7, article 195. doi: 10.3389/fnagi.2015.00195.

Tom N. Tombaugh and Nancy J. McIntyre. 1992. The Mini-Mental State Examination: A Comprehensive Review. *Progress in Geriatrics.* Vol. 40:9, pp. 922–935 doi: 10.1111/j.1532-5415.1992.tb01992.xView/.

Laszló Tóth, Gábor Gosztolya, Veronika Vincze, Ildiko Hoffmann, Gráta Szatloczki, Edit Biro, Fruzsina Zsura, Magdolna Pákáski and János Kálmán. 2015. *Automatic Detection of Mild Cognitive Impairment from Spontaneous Speech using ASR.* Proceedings of Interspeech. Dresden, Germany.

Susanne Trauzettel-Klosinski, Klaus Dietz and the IReST Study Group. 2012. Standardized Assessment of Reading Performance: The New International Reading Speed Texts IReST. *Investigative Ophthalmology & Visual Science.* Vol. 53:9. The Association for Research in Vision and Ophthalmology, Inc.

Olof Tyche. 2001. *Subtila språkstörningar hos patienter med diagnosen MCI (Mild Cognitive Impairment) Del I: utifrån den tematiska bilden "Kakstölden".* KI, Sweden (In Swedish).

Anders Wallin et al. 2016. The Gothenburg MCI study: Design and distribution of Alzheimer's disease and subcortical vascular disease diagnoses from baseline to 6-year follow-up. *J Cereb Blood Flow Metab.* 36(1):114-31.

Caroline Williams, Andrew Thwaites , Paula Buttery, Jeroen Geertzen, Billi Randall, Meredith Shafto, Barry Devereux and Lorraine Tylera. 2010. *The Cambridge Cookie-Theft Corpus: A Corpus of Directed and Spontaneous Speech of Brain-Damaged Patients and Healthy Individuals.* 7th Language Resources and Evaluation Conference, LREC. Pp. 2824-2839. Malta.

Victoria Yaneva, Irina Temnikova and Ruslan Mitkov. 2016. *Corpus of Text Data and Gaze Fixations from Autistic and Non-autistic Adults.* Proceedings of the 10th Language Resources and Evaluation Conference (LREC), pp. 480-487. Portorož, Slovenia.

Maria Yancheva, Kathleen Fraser and Frank Rudzicz. 2015. *Using linguistic features longitudinally to predict clinical scores for Alzheimer's disease and related dementias.* 6th Workshop on Speech and Language Processing for Assistive Technologies (SLPAT), pp. 134–139, Dresden, Germany.

Gustaf Öqvist Seimyr. 2010. *Swedish IReST translation.* The Bernadotte Laboratory, Karolinska institute, Sweden.

Analysis of Anxious Word Usage on Online Health Forums

Nicolas Rey-Villamizar
Dept. of Computer Science
University of Houston
Houston, TX 77204-3010
`nrey@uh.edu`

Prasha Shrestha
Dept. of Computer Science
University of Houston
Houston, TX 77204-3010
`pshrestha3@uh.edu`

Farig Sadeque
School of Information
University of Arizona
Tucson, AZ 85721
`farig@email.arizona.edu`

Steven Bethard
School of Information
University of Arizona
Tucson, AZ 85721
`bethard@email.arizona.edu`

Ted Pedersen
Dept. of Computer Science
University of Minnesota, Duluth
Duluth, MN 55812-3036
`tpederse@d.umn.edu`

Arjun Mukherjee, Thamar Solorio
Dept. of Computer Science
University of Houston
Houston, TX 77204-3010
`arjun@cs.uh.edu, solorio@cs.uh.edu`

Abstract

Online health communities and support groups are a valuable source of information for users suffering from a physical or mental illness. Users turn to these forums for moral support or advice on specific conditions, symptoms, or side effects of medications. This paper describes and studies the linguistic patterns of a community of support forum users over time focused on the used of anxious related words. We introduce a methodology to identify groups of individuals exhibiting linguistic patterns associated with anxiety and the correlations between this linguistic pattern and other word usage. We find some evidence that participation in these groups does yield positive effects on their users by reducing the frequency of anxious related word used over time.

1 Introduction

How people behave within a given community is an important question, especially in the context of health support. The advancement of technology has complemented the classic in-person health support forums to a growing and vibrant online community. Social media research has indicated that individuals psychological states and social support status relating to health and well-being may be deduced via analysis of language and conversational patterns (Tamersoy et al., 2015). In the offline world, some psychological studies of people's behavior have shown correlation of different sociological dimensions such as sadness and anger over the time course of a breakup process (Sbarra, 2006). We want to study these kinds of correlations in online support forums.

In this paper we focus our attention on the analysis of the users who participated in the Daily-Strength forums[1], and we propose a methodology to study the users' behaviors by analyzing the linguistic characteristics of their posts. Researchers have shown that a large and increasing number of people are going online for medical information and advice (Fox and Duggan, 2013). We focus our study on the usage of words related to anxiety. This is an important area of interest for us given that previous research has shown that in some age groups up to 33.7% of participants are diagnosed with some type of Anxiety Disorder (Chou, 2010).

Some researchers have found a correlation be-

---

[1] https://www.dailystrength.org/

37

Proceedings of the Seventh International Workshop on Health Text Mining and Information Analysis (LOUHI), pages 37–42,
Austin, TX, November 5, 2016. ©2016 Association for Computational Linguistics

tween the usage of words related to anxiety with daily negative emotions (Tov et al., 2013). Applying our proposed framework, we found that the usage of words related to anxiety by active users in an online health support group has a steady decrease over the course of a user's involvement in the community and we theorize their daily negative emotion reduces as well. Our proposed framework can be easily extended to other related conditions such as depression or eating disorders. In general, we believe that sociolinguistic characteristics in online health support forums is an exciting topic that can shed additional light on human behavior and on the design of social media systems.

2   Related Work

Online Health Forums and social media: Online health communities are a rich source of data for the research community as a whole. Some researchers have studied the potential and limitation of such data and how it can augment existing public health capabilities and enable new ones (Dredze, 2012). One of the major concerns is the credibility of the information. Other researchers have studied how to automatically establish the credibility of the user generated medical statements by analyzing linguistic clues (Mukherjee et al., 2014). Other researchers have focused on understanding abstinence from tobacco or alcohol use (Tamersoy et al., 2015) and on how to find early indications of Adverse Drug Reactions from online healthcare forums (Sampathkumar et al., 2014). In the online world, several of the largest online health community websites are: MedHelp (`www.medhelp.org`), Patients-LikeMe (`www.patientslikeme.org`), and Daily-Strength (`www.dailystrength.org`)

Sociolinguistic patterns in social media: Social media is very appealing to the study of sociolinguistic analysis of the users. In particular one of the main concerns is how much information is actually posted by the users to justify the study of such textual data. In (Park, 2012) the authors found evidence that people post about their treatment on social media. Some researchers have shown the predictive power of studying linguistic patterns of social media users in order to predict depression (De Choudhury and Gamon, 2013). Furthermore, other researchers

have studied insights about diseases, such as analyzing symptoms and medication usage and have found a strong correlation with public health data (Passarella, 2011). Ofoghi et al. (2016) have created an emotion classification of microblog content in order to study the public mood and effectively utilize it as an early warning system for epidemic outbreak. Also, they analyzed the emotions in microblog content after outbreaks to validate their approach. Finally, Aman and Szpakowicz (2007) describe an emotion annotation task and study how the inter-annotator agreement. They show how difficult is the emotion annotation task, the inter-annotator agreement ranges between 0.6 to 0.79.

Anxiety disorders: Researchers estimate the percentage of adults with anxiety disorders to vary from 3.2% (Fuentes and Cox, 1997) up to 14.2% (Norton et al., 2012). Other researchers have suggested that up to 33% of the general population will develop "clinical significant anxiety disorder" at some time in their life (Barlow et al., 2002). Anxiety disorders are commonly associated with medical conditions such as thyroid disease, asthma and heart disease (Diala and Muntaner, 2003). Also, some conditions such as coronary heart disease, hypertension, and hypoglycemia can be worsened through anxiety (Hersen and Van Hasselt, 1992).

3   Dataset Description

We collected data from Daily-Strength, one of the largest online support groups with more than 500 active groups based on the physical and mental conditions of its users. Daily-Strength allows users to create profiles, maintain friends, and join various condition-related support groups. It serves as a resource for patients to connect with others who have similar conditions. Users in these support groups can either *create* a new thread on a new topic[2], or *reply*[3] to a thread that someone else has created.

In the current study, we selected the support groups that had the most vibrant communities based on the number of unique users, and the number of unique threads. We focused on support groups with more than 1,000 different users and more than 200

---

[2]The topics are curated by the system administrators.

[3]The website does not distinguish between a reply to a main thread and a reply to a reply.

Table 1: Dataset statistics

Characteristic	Value
Number of support groups	93
Number of unique users	193,354
Number of unique posts	10,612,830

original threads. At the time of our data crawl, 93 groups fulfilled this selection constraint. Some of the most active support groups in this list include: Acne, ADHD (Attention Deficit Hyperactivity Disorder), Alcoholism, Asthma, Back Pain, Bipolar Disorder, Bone Cancer, COPD (Chronic Obstructive Pulmonary Disease), and Fibromyalgia. We crawled all of the original posts, thread initiations, and all the user replies for these support groups from the earliest available post until March 25, 2015. The posts and replies were downloaded as HTML files, one per thread, where each thread contains an initial post and zero or more replies. We filtered out posts from administrators of the website since they do not reflect the user's activities but just general guidelines or advice for the users.

Researchers have shown that a large and increasing number of people are going online for medical information and advice (Fox and Duggan, 2013). We want to base our analysis on a group of users who are consistently involved in the forum. Moreover, since we're interested in exploring if participation in the forum has any effects on its users and if these effects are reflected in linguistic patterns, it's important to analyze data from user posts across a significant period of time. We believe one year of activity will fulfill this purpose. Thus in subsequent analyses we filtered by users whose first and last post are at least one year apart, and who posted at least 50 posts during that interval. This filtering reduced the total number of users to approximately 10,000 users, still a significant number for our studies.

4   Analysis framework

We defined user behavioral dimensions (BDs) based on the word list provided by the Linguistic Inquiry and Word Count (LIWC) lexicon (Tausczik and Pennebaker, 2010). Each BD is defined as the corresponding word list from the LIWC lexicon. The LIWC contains 4,500 words and word stems, and each word or word stem defines one or more word categories or subdictionaries. This is a well developed tool aimed at revealing our thoughts, feelings, personality, and motivations based on our word usage. We focused on the anxious word list, but our analysis can be easily extended to any of the other LIWC list. Based on the emission of words in the linguistic dimensions in LIWC we created what we call user behavioral dimensions (BD). In order to study how the users of anxious words relate to other BDs defined by the LIWC we present in subsection 4.3 the correlation with six other BDs.

4.1   User's sub-population selection

The first part of our framework consists of creating a methodology to study a specific user's behavioral dimension. This methodology will help us find sub-populations corresponding to a particular BD. For this we propose metric for each of the BDs. We quantify each user BD according to:

$$BD_i(u) = \log\left(\frac{1}{|posts(u)|} \sum_{p \in posts(u)} \frac{|words_{BD_i}(p)|}{|words(p)|}\right)$$

where $i \in \{1, \ldots, N\}$ indexes the BDs, N is the number of different BDs, $posts(u)$ is the list of posts for user u, $words(p)$ is the list of words in post p, and $words_{BD_i}(p)$ is the list of words in post p that were on the list BD_i. In essence, we measure the average fraction of words from the list BD_i across the posts of a user. Since these fractions are less than 1 and we are using a logarithmic scale, $BD_i(u)$ will always have a negative magnitude.

We propose to base our analysis on the study of the extreme populations. To do that, we create a histogram which allows us to cluster users at the tails of the distribution. We create 4 clusters per condition out of this histogram. The lower 15% and 30% users; the top 30% and 15%. Figure 1 shows the generated histogram for the anxious word list from the LIWC. In order to select each population we sort the users according to the quantified values of the BD and then select the upper/lower part of the users.

4.2   Anxious word usage

We present the methodology used to analyze the BD corresponding to anxious words. The other BDs can be analyzed in an analogous way. The methodology we developed to analyze anxious word usage

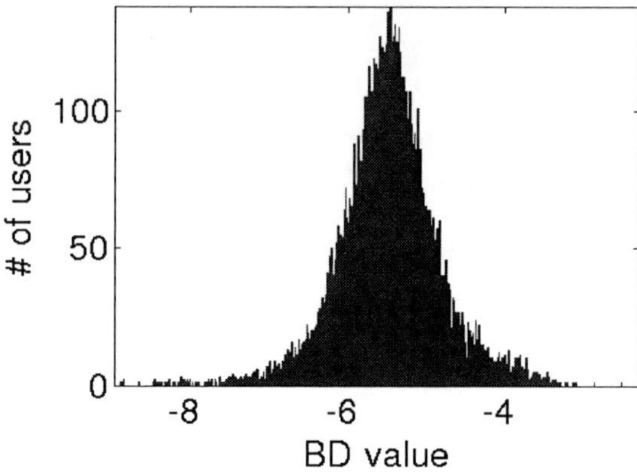

Figure 1: Sample usage of anxious related words.

is as follows. We first group all the posts based on the month when the user posted. All posts from the users during their first month, second month, third month, and so on. We then analyzed the usage pattern of anxious words over time. Figure 2 shows that the anxiousness of the users decreases in a constant manner over the course of their active involvement in the forum. This is in agreement with the effect of off-line support groups (Cain et al., 1986).

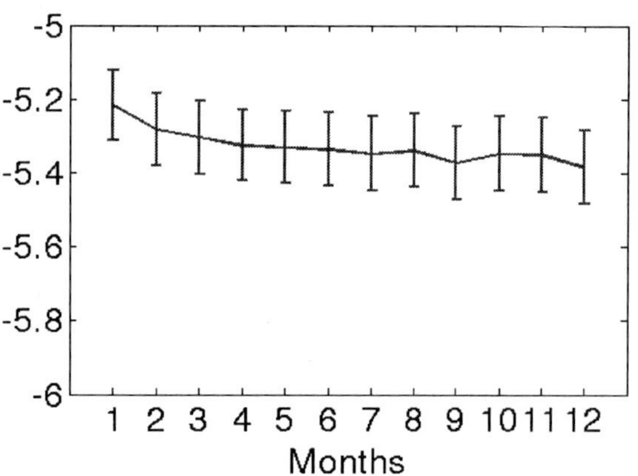

Figure 2: Single BD, anxious word usage.

4.3 Correlation of anxious words with other word usage

In this part we focus on the analysis of BDs that correlate the most with anxious word usage. We ranked the BDs based on the correlation with anxious words usage. We used the Pearson coefficient correlation. In descending order the most correlated BDs are: a) anger, b) self-pronouns, c) death, d) money, e)

present, and f) body. We present the results of all the correlations for completeness and describe the ones we found most interesting. For each of the clusters of users we identified on the anxious word usage, we plot the secondary BD score over time in Figure 3.

Figure 3 panels (a), (b), (c), and (e), shows that the usage of anger, self, death, and present words also decreases over time for all the low and high heavy users of anxious words. However from panels (d) and (f) we see that there is an increase of that particular word usage for some of the groups and a decrease from other groups.

Figure 3 (c) illustrates this for the case of anchor usage of anxious words and the figure shows the usage pattern of death words by those groups of users. We can see that users who used more anxious words (magenta dotted and black dash-dot) consistently used more death words over time than users who used fewer anxious words (red line and blue dashed). We theorize that this pattern can be related to suicidal topics similar to what other researchers have reported for depressed and anxious patients (Pompili et al., 2012)

From Figure 3 panel (d) we can see a difference between the amount of money related words used by people who use anxious words. The groups of users who use less anxious words (red line and blue dashed) tend to use more money related words, whereas the groups of users who use more anxious related words (magenta dotted and black dash-dot) tend to use less money words. In general researchers have linked people being tight with money as having more negative emotions such as been more anxious (McClure, 1984), however to the best of our knowledge this is the first study of such relation into a health support context where money is probably not directly related to how wealthy or not an individual is. From Figure 3 panel (e) we can see that people who use more anxious words (magenta dotted and black dash-dot) tend to use more present words, whereas people who use less anxious words (red line and blue dashed) use less present words. Some researchers have previously linked Defensive Pessimism people with being anxious in the present (Norem and Smith, 2006), we theorize that it can be a general pattern of people participating in online support forums.

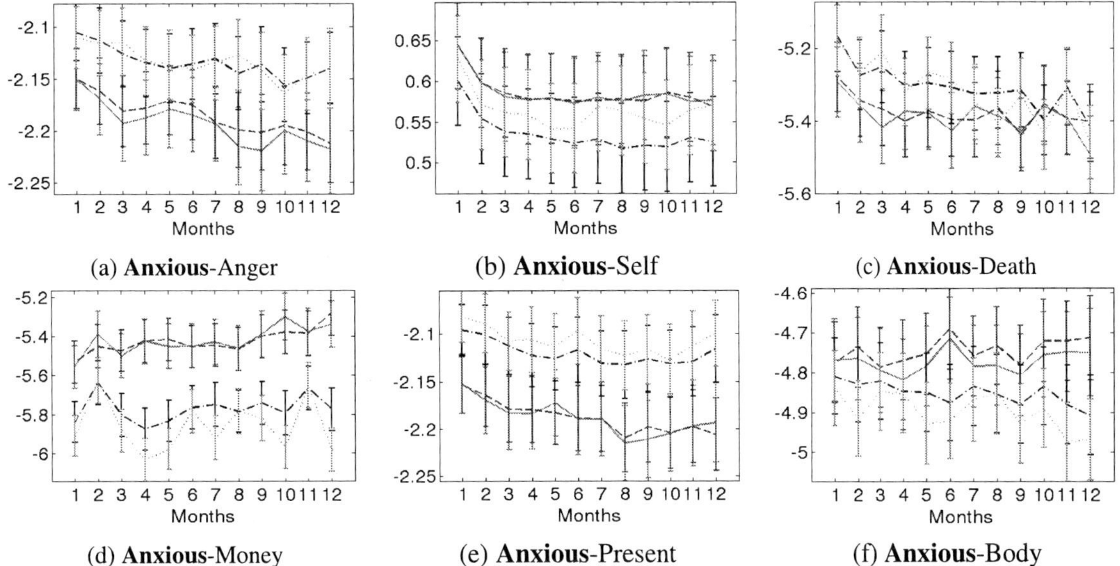

(a) Anxious-Anger   (b) Anxious-Self   (c) Anxious-Death

(d) Anxious-Money   (e) Anxious-Present   (f) Anxious-Body

Figure 3: Correlation of the different BDs. Anchor BD anxious word usage (in boldface). Red line (-), blue line (–), black (-.), and magenta (..) corresponds to the lower 15%, 30% and upper 30%, 15% of the anxious word usage distribution.

5 Conclusion and Future Work

Similar to what other researchers (Cain et al., 1986) have shown for off-line support groups, based on our proposed framework for analyzing linguistic patterns of users of online support groups we conclude that the anxiety levels of patients involved in support groups lowers over time. We also conclude that anxiety levels are not directly related to money related talks in online support forums participants.

In this paper we have presented the correlation between Anxiety and BDs that we think are more interesting to study and the ones which are more relevant given the literature on Anxiety. However, a more detailed and robust method is been developed in order to rank the most relevant BDs which exhibit significant correlation over time.

References

Saima Aman and Stan Szpakowicz. 2007. Identifying expressions of emotion in text. In *International Conference on Text, Speech and Dialogue*, pages 196–205. Springer.

David H Barlow, Susan D Raffa, and Elizabeth M Cohen. 2002. *Psychosocial treatments for panic disorders, phobias, and generalized anxiety disorder*, volume 2.

Eileen N. Cain, Ernest I. Kohorn, Donald M. Quinlan, Kate Latimer, and Peter E. Schwartz. 1986. Psychosocial benefits of a cancer support group. *Cancer*, 57(1):183–189.

Kee-Lee Chou. 2010. Panic disorder in older adults: evidence from the national epidemiologic survey on alcohol and related conditions. *International journal of geriatric psychiatry*, 25(1-3):822–832.

Munmun De Choudhury and Michael Gamon. 2013. Predicting depression via social media. *Proceedings of the Seventh International AAAI Conference on Weblogs and Social Media*, 2:128–137.

Chamberlain C. Diala and Carles Muntaner. 2003. Mood and anxiety disorders among rural, urban, and metropolitan residents in the United States. *Community Mental Health Journal*, 39(3):239–252.

Mark Dredze. 2012. How social media will change public health. *IEEE Intelligent Systems*, 27(4):81–84.

Susannah Fox and Maeve Duggan. 2013. Pew Internet & American Life Project. online http://www.pewinternet.org/2013/01/15/health-online-2013/.

Karina Fuentes and Brian J Cox. 1997. Prevalence of anxiety disorders in elderly adults: A critical analysis. *Journal of Behavior Therapy and Experimental Psychiatry*, 7916(4):269–279.

Michel Hersen and Vincent B. Van Hasselt. 1992. Behavioral assessment and treatment of anxiety in the elderly. *Clinical Psychology Review*, 12(1984):619–640.

Robert F. McClure. 1984. The relationship between money attitudes and overall pathology. *Psychology: A Journal of Human Behavior*, 21(1):4–6.

Subhabrata Mukherjee, Gerhard Weikum, and Cristian Danescu-Niculescu-Mizil. 2014. People on drugs: credibility of user statements in health communities. *KDD '14: Proceedings of the 20th ACM SIGKDD international conference on Knowledge discovery and data mining.*

JK. Norem and S. Smith. 2006. Defensive pessimism: positive past, anxious present, and pessimistic future. In *Judgements over time: The interplay of thoughts, feelings, and behaviors*, pages 34–46.

Joanna Norton, Maria Laure Ancelin, Rob Stewart, Claudine Berr, Karen Ritchie, and Isabelle Carriere. 2012. Anxiety symptoms and disorder predict activity limitations in the elderly. *Journal of Affective Disorders*, 141:276–285.

Bahadorreza Ofoghi, Meghan Mann, and Karin Verspoor. 2016. Towards early discovery of salient health threats: A social media emotion classification technique. In *Pacific Symposium on Biocomputing. Pacific Symposium on Biocomputing*, volume 21, page 504.

Minsu Park. 2012. Depressive moods of users portrayed in twitter. *ACM SIGKDD Workshop on Healthcare Informatics (HI-KDD)*, page 18.

Ralph J Passarella. 2011. You are what you tweet : Analyzing twitter for public health. *In Empirical Natural Language Processing Conference (EMNLP).*, 56(8):1–12.

Maurizio Pompili, Marco Innamorati, Zoltan Rihmer, Xenia Gonda, Gianluca Serafini, Hagop Akiskal, Mario Amore, Cinzia Niolu, Leo Sher, Roberto Tatarelli, Giulio Perugi, and Paolo Girardi. 2012. Cyclothymic-depressive-anxious temperament pattern is related to suicide risk in 346 patients with major mood disorders. *Journal of affective disorders*, 136(3):405–11.

Hariprasad Sampathkumar, Xue-wen Chen, and Bo Luo. 2014. Mining adverse drug reactions from online healthcare forums using hidden Markov model. *BMC medical informatics and decision making*, 14(1):91.

David A. Sbarra. 2006. Predicting the onset of emotional recovery following nonmarital relationship dissolution: Survival analyses of sadness and anger. *Personality and Social Psychology Bulletin 32*, 3:298–312.

Acar Tamersoy, Munmun De Choudhury, and Duen Horng Chau. 2015. Characterizing smoking and drinking abstinence from social media. In *Proceedings of the 26th ACM Conference on Hypertext & Social Media*, pages 139–148. ACM.

Yla R. Tausczik and James W. Pennebaker. 2010. The psychological meaning of words: LIWC and computerized text analysis methods. *Journal of Language and Social Psychology*, 29(1):24–54.

William Tov, Kok Leong Ng, Han Lin, and Lin Qiu. 2013. Detecting well-being via computerized content analysis of brief diary entries. *Psychological assessment*, 25(4):1069–78.

Retrofitting Word Vectors of MeSH Terms to Improve Semantic Similarity Measures

Zhiguo Yu, MS, Trevor Cohen, MBChB, PhD
Elmer V. Bernstam, MD, MSE, Todd R. Johnson, PhD
The University of Texas Health Science Center at Houston, Houston, TX
`Zhiguo.yu@uth.tmc.edu`

Byron C. Wallace, PhD
College of Computer and Information Science, Northeastern University, Boston, MA

Abstract

Estimation of the semantic relatedness between biomedical concepts has utility for many informatics applications. Automated methods fall into two broad categories: methods based on distributional statistics drawn from text corpora, and methods based on the structure of existing knowledge resources. In the former case, taxonomic structure is disregarded. In the latter, semantically relevant empirical information is not considered. In this paper, we present a method that retrofits the context vector representation of MeSH terms by using additional linkage information from UMLS/MeSH hierarchy such that linked concepts have similar vector representations. We evaluated the method relative to previously published physician and coder's ratings on sets of MeSH terms. Our experimental results demonstrate that the retrofitted word vector measures obtain a higher correlation with physician judgments. The results also demonstrate a clear improvement on the correlation with experts' ratings from the retrofitted vector representation in comparison to the vector representation without retrofitting.

1 Introduction

Groups of semantically similar concepts and terms are known to improve the retrieval (Rada et al., 1989) and clustering (Lin et al., 2007) of biomedical and clinical documents, and the development of biomedical terminologies and ontologies (Bodenreider and Burgun, 2004). However, automated estimation of semantic similarity remains a challenge. Most semantic similarity measures leverage the structure of an ontology or taxonomy (e.g. WordNet, Unified Medical Language System (UMLS)/Medical Subject Headings (MeSH)) to calculate, for example, the shortest path information between concept nodes (Pedersen et al., 2007; Caviedes and Cimino, 2004). Vector representations based on a co-occurrence matrix from a corpus has also been used to calculate the relatedness between concepts (Pedersen et al., 2007; Pedersen et al., 2004). Others use information content (IC) to estimate the semantic similarity and relatedness between two concepts, which incorporate the probability of the concept occurring in a corpus (Caviedes and Cimino, 2004; Ciaramita et al., 2008; Turney, 2005). Some topic modeling techniques (Blei et al., 2003; Yu et al., 2013) have also been applied to integrate the automatically generated themes (topics) from a specific corpus to the controlled vocabulary that indexed within this corpus to help improve the document retrieval and clustering performances (Yu et al., 2016).

In this paper, we introduce a new semantic similarity measure utilizing both vector space word representations and a biomedical taxonomy (UMLS/MeSH) to determine the degree of semantic similarity between pairs of concepts. For two concepts, we first learn their vector space word representations from distributional information of words in a large domain-relevant corpus. Although such vectors are semantically informative, they disregard the valuable information contained in semantic lexicons such as WordNet, FrameNet, and the Paraphrase Database. In 2014, Faruqui, et al. (Faruqui et al., 2014a) developed a "retrofitting"

Proceedings of the Seventh International Workshop on Health Text Mining and Information Analysis (LOUHI), pages 43–51,
Austin, TX, November 5, 2016. ©2016 Association for Computational Linguistics

method that addresses this limitation by incorporating information from such semantic lexicons into word vector representations, such that semantically linked words will have similar vector representations. We applied this technique to word vector representations of UMLS/MeSH concepts in an effort to improve their quality. We evaluated the method relative to previously published human expert similarity ratings of a Physician and Coder on sets of MeSH terms. Our experimental results demonstrate that the retrofitted word vector similarity measures have a higher correlation with Physician (but not Coder) judgments, compared with other existing techniques. The results also demonstrate a clear improvement on the correlation with experts' ratings from the retrofitted vector representation to the vector representation without retrofitting.

2   Related Work

There are two major classes of semantic similarity measurement methods. The most common class uses an ontology or taxonomy to calculate the shortest path between two concepts. Rada, et al. (Rada et al., 1989) introduces the measure of conceptual distance to quantify the similarity between concepts in the UMLS. Wu and Palmer (Wu and Palmer, 1994) extend this measure by calculating the length of shortest path between two concepts that connects the concepts through their least common subsumer (LCS). The LCS is the most specific ancestor shared by two concepts. In 2005, Nguyen and Al-Mubaid (Nguyen and Al-Mubaid, 2006) proposed a new path-based measure using $is - a$ relation in MeSH. They incorporate both the depth and LCS in their measure. In their results, they compared with the measures introduced by Leacock & Chodorow (Leacock and Chodorow, 1998), Wu & Palmer (Wu and Palmer, 1994), and the Path measure. Batet, et al. (Batet et al., 2011) introduce a measure that incorporates the common concepts shared between the two concepts and their LCS. Recently, McInnes, et al. (McInnes et al., 2014) introduced U-path measure using undirected path to determine the degree of semantic similarity between two concepts in a dense taxonomy with multiple inheritance. In 2009, McInnes, et al. (McInnes et al., 2009) presented a UMLS-Similarity tool which contains five semantic similarity measures proposed by Rada, et al. (Rada et al., 1989) , Wu & Palmer (Wu and Palmer, 1994), Leacock & Chodorow (Leacock and Chodorow, 1998), and Nguyen & Al-Mubaid (Nguyen and Al-Mubaid, 2006), and the Path measure.

The second class of techniques uses training corpora and information content (IC) to estimate the semantic similarity between two concepts. IC measures the specificity of a concept in a hierarchy. The IC-based measures account for the probability of the concept occurring in a corpus. A concept with a high IC value is more specific to a topic than one with a low IC value. Resnik (Resnik, 1995), Jiang & Conrath (Jiang and Conrath, 1997) and Lin (Lin, 1998), all have published works on the IC-based similarity measures. Resnik (Resnik, 1995) measures the similarity between two concepts by finding the IC of the LCS of the two concepts. Jiang & Conrath (Jiang and Conrath, 1997) and Lin (Lin, 1998) extended Resnik's IC-based measure by incorporating the IC of the individual concepts. Jiang & Conrath measure similarity by finding the IC of each individual concept and of the LCS of them. However, Lin's measure is similar to that of Wu & Palmer (Wu and Palmer, 1994), where depth is replaced by information content.

Context vector metrics based on distributional statistics have also been used to calculate semantic similarity (Patwardhan, 2006; Patwardhan, 2003). By building co-occurrence vectors that represent the contextual profile of concepts, the relatedness between concepts can then be calculated using cosine similarity between vectors corresponding to two given concepts (Pedersen et al., 2007).

Though IC-based measures do draw upon distributional information, this is used in a very restricted way to determine the specificity of a concept. Context vector metric-based distributional statistics do not have such limitations on the use of distributional information. However, the taxonomic structure is not taken into account in distributional methods. "Correlation with human pairwise judgment" evaluation is widely used in computational linguistics. There are a number of evaluation sets exist in the biomedical domain. 'MayoSRS', developed by

Pakhomov, et al. (Pakhomov et al., 2011), consists of 101 clinical term pairs whose relatedness was determined by nine medical coders and three physicians from the Mayo Clinic. In this paper, we used 'MiniMayoSRS,' a subset of 'MayoSRS.' The average correlation between physicians is 0.68. The average correlation between medical coders is 0.78 (Pedersen et al., 2007). 'UMNSRS', developed by Pakhomov, et al. (Pakhomov et al., 2010), consists of 725 clinical term pairs whose semantic similarity and relatedness was determined independently by four medical residents from the University of Minnesota Medical School.

3   Method

In this section, we provide a brief description of the method used for retrofitting word vector to semantic lexicons, present the design of our work flow, describe the test data and the semantic lexicon we created , and also present the evaluation measures we used.

3.1   Retrofitting Word Vector to Semantic Lexicons

Vector space word representations are a critical component of many natural language processing systems. It is common to represent words as discrete indices in a vocabulary, but this fails to capture the rich relational structure of the human semantic lexicon (Maas et al., 2011). Retrofitting is a simple and effective method to improve word vectors using word relation knowledge found in semantic lexicons. It is used as a post-processing step to improve vector quality (Faruqui et al., 2014a).

Figure 1 shows a small word graph example with edges connecting semantically related words. The words, *cancer, tumor, neoplasm, sarcoma*, and *swelling*, are similar words to each other in a lexical knowledge resource. Grey nodes are observed word vectors built from the corpus, which are independent of each other. White nodes are inferred word vectors, waiting to be retrofitted. The edge between each pair of white nodes means they are similar words to each other. The inferred word vector (e.g., *q_tumor*) is expected to be close to its cor-

responding observed word vector (e.g., $\hat{q}\_tumor$) and close to its synonym neighbors (e.g., *q_cancer* and *q_neoplasm*). The objective is to minimize the following:

$$\Psi(Q) = \sum_{i=1}^{n} [\alpha_i \|q_i - \hat{q_i}\|^2 + \sum_{(i,j) \in E} \beta_{ij} \|q_i - q_j\|^2] \tag{1}$$

where α and β values control the relative strengths of associations, Q is the retrofitted vectors, and $(i, j) \in E$ means there is an edge between node q_i and q_j. Ψ is convex in Q. An efficient iterative updating method is used to find this convex. First, retrofitted vectors in Q are initialized to be equal to the observed vectors. The next step is to take the first derivative of Ψ with respect to q_i vector and use the following to update it online.

$$q_i = \frac{\sum_{j:(i,j) \in E} \beta_{ij} q_j + \alpha_i \hat{q_i}}{\sum_{j:(i,j) \in E} \beta_{ij} + \alpha_i} \tag{2}$$

It takes approximately 10 iterations to converge to the difference in Euclidean distance of adjacent nodes of less than 0.01 in practice. An implementation of this algorithm has been published online by the authors (Faruqui et al., 2014b). We used this implementation in the current work.

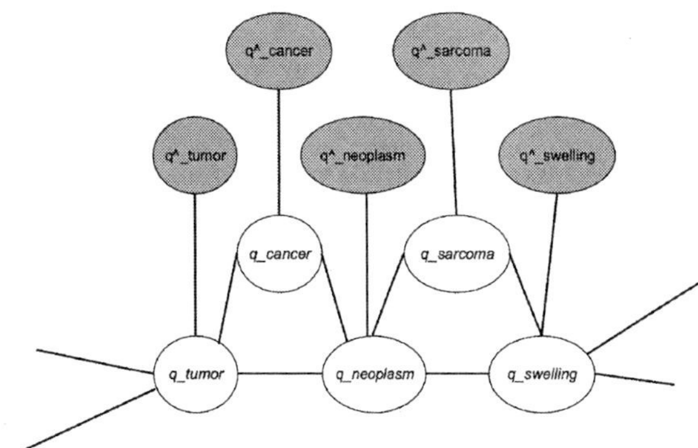

Figure 1: Word graph with edges between related words, observed (grey node), inferred (white node).

3.2   Work Flow

Our work flow is presented in Figure 2. The input is a pair of concepts. The output is a similarity score. The next step after *input data* is to *fetch relevant*

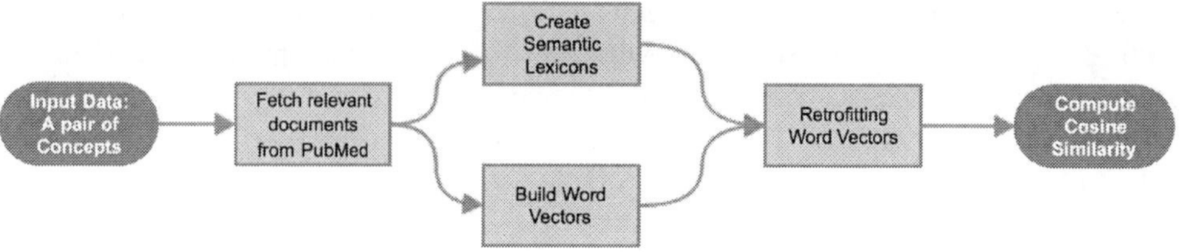

Figure 2: Work flow.

documents from PubMed. In our test data, each concept is mapped to MeSH term(s) (Please see details in the paragraph *Test Data* of this section.) We then randomly fetch 1000 citations indexed with those MeSH term(s) from PubMed. In the *Build Word Vectors* step, we build each MeSH term a word vector using the approach described in (Yu et al., 2016). We use titles and abstracts of returned citations and only select those MeSH terms indexed in more than 100 citations as our candidate semantic lexicon. The main MeSH term mapped from the input concept is indexed in all 1000 citations. The retrofitted vector quality suffers if we take into account MeSH terms that appear in a small number of citations. For each selected MeSH term, we collect all the words from the citations indexed with that MeSH term. After removing the stop words, We use *tf-idf* (Equation 3) to weight the remaining words and then normalize the weights so that they sum to one. In *Create Semantic Lexicons*, we use both the UMLS-similarity tool developed by McInnes, et al. (McInnes et al., 2009) and the MeSH tree structure as the source from which it estimates semantic relatedness. For details see the paragraph *Semantic Lexicons* in this section. *Retrofitting Word Vectors* retrofits the word vectors using the created semantic lexicons to generate new word vectors. We then calculate cosine similarity (Equation 4) based on the concepts pair's new word vectors. On account of the stochastic nature of the literature sampling, we test each pair of concepts five times and average performance over these five times as its final similarity score.

$$tf\text{-}idf_{w,d} = tf_{w,d} * log\frac{N}{df_{w,D}} \quad (3)$$

where $tf_{w,d}$ is the term frequency of word w in document d, $df_{w,D}$ is the document frequency that word w appears in all documents D, and N is the total number of documents.

$$cos(\theta) = \frac{A \cdot B}{\|A\| \cdot \|B\|} = \frac{\sum_{i=1}^{n} A_i B_i}{\sum_{i=1}^{n} A_i^2 \sum_{i=1}^{n} B_i^2} \quad (4)$$

where A_i and B_i are components of vector A and B respectively.

3.3 Test Data

We used the set of 30 concept pairs from Pedersen, Pakhomov, and Patwardhan (2005) (Pedersen et al., 2007), which was annotated by 3 physicians and 9 medical index coders. Each pair was annotated on a 4 point scale: *"practically synonymous, related, marginally, and unrelated"*. Table 1 displays the details of these concepts pairs along with both ratings.

Neguyen and Al-Mubaid use 25 out of the 30 pairs of terms in the dataset. 5 pairs of terms (highlighted in both table 1 and table 2) were excluded because they did not exist in MeSH version 2006. To make it comparable with their results, we also use these 25 pairs of terms. The mappings of the terms to MeSH terms were obtained firstly by using the online MetaMap tool (Aronson and Lang, 2010). Then we used the MeSH browser 2016 (MeSH, 2016) to get the most updated MeSH terms.

3.4 Semantic Lexicons

We tested two semantic lexicons in our experiments. The first is from the results of McInnes, et al.'s UMLS-Similarity tool (McInnes et al., 2009). UMLS-Similarity contains five semantic similarity measures proposed by Rada, et al. (Rada et al., 1989), Wu & Palmer (wup) (Wu and Palmer, 1994), Leacock & Chodorow (lch) (Leacock and Chodorow, 1998), and Nguyen & Al-Mubaid (nam) (Nguyen and Al-Mubaid, 2006), and the Path measure. Leacock & Chodorow's measure achieved best performance among these five semantic similarity measures. In our experiment, we used this mea-

Term 1	Term 2	Physicians	Coders
Renal failure	Kidney failure	4.0000	4.0000
Heart	Myocardium	3.3333	3.0000
Stroke	Infarct	3.0000	2.7778
Abortion	miscarriage	3.0000	3.3333
Delusion	Schizophrenia	3.0000	2.2222
Congestive heart failure	Pulmonary edema	3.0000	1.4444
Metastasis	Adenocarcinoma	2.6667	1.7778
Calcification	Stenosis	2.6667	2.0000
Diarrhea	**Stomach cramps**	2.3333	1.3333
Mitral stenosis	Atrial fibrillation	2.3333	1.3333
Chronic obstructive pulmonary disease	**Lung infiltrates**	2.3333	1.8889
Rheumatoid arthritis	Lupus	2.0000	1.1111
Brain tumor	Intracranial hemorrhage	2.0000	1.3333
Carpal tunnel syndrome	Osteoarthritis	2.0000	1.1111
Diabetes mellitus	Hypertension	2.0000	1.0000
Acne	Syringe	2.0000	1.0000
Antibiotic	Allergy	1.6667	1.2222
Cortisone	Total knee replacement	1.6667	1.0000
Pulmonary embolus	**Myocardial infarction**	1.6667	1.2222
Pulmonary Fibrosis	Lung Cancer	1.6667	1.4444
Cholangiocarcinoma	Colonoscopy	1.3333	1.0000
Lymphoid hyperplasia	Laryngeal Cancer	1.3333	1.0000
Multiple Sclerosis	Psychosis	1.0000	1.0000
Appendicitis	Osteoporosis	1.0000	1.0000
Rectal polyp	**Aorta**	1.0000	1.0000
Xerostomia	Alcoholic cirrhosis	1.0000	1.0000
Peptic ulcer disease	Myopia	1.0000	1.0000
Depression	Cellulitis	1.0000	1.0000
Varicose vein	**Entire knee meniscus**	1.0000	1.0000
Hyperlipidemia	Metastasis	1.0000	1.0000

Table 1: Test set of 30 medical term pairs sorted in the order of the averaged physician' scores.

sure in UMLS-Similarity to calculate the similarity score between each selected MeSH term and the main MeSH term. We calculated the average of all these scores as the threshold. We then chose those MeSH terms whose scores are over this threshold as the main MeSH term's semantic lexicon terms. The second semantic lexicon is constructed using MeSH tree structure information. For each main MeSH term, we chose its parents and child terms from the MeSH tree as its lexicon terms.

3.5 Evaluation

In our experiment, we used three types of vector representations to calculate the semantic similarity: MeSH term word vectors without retrofitting; MeSH term word vectors retrofitted with UMLS-Similarity results; and MeSH term word vectors retrofitted using the MeSH tree structure. We rank the 25 pairs of terms based on similarity scores and calculate the correlation between our rankings and the Physician and Coder judgments using the Spearman rank correlation coefficient. We compare our correlation results with those reported by Nguyen, et al. (Nguyen and Al-Mubaid, 2006) and

Term 1	Term 2	Word Vector	Retrofitted with UMLS-Similarity Results	Retrofitted with MeSH Tree Structure
Renal failure	Kidney failure	1.00	1.00	1.00
Heart	Myocardium	0.86	0.85	0.86
Stroke	Infarct	0.70	0.71	0.70
Abortion	miscarriage	0.79	0.74	0.76
Delusion	Schizophrenia	0.81	0.83	0.81
Congestive heart failure	Pulmonary edema	0.73	0.72	0.73
Metastasis	Adenocarcinoma	0.88	0.84	0.83
Calcification	Stenosis	0.47	0.46	0.47
Diarrhea	**Stomach cramps**	N/A	N/A	N/A
Mitral stenosis	Atrial fibrillation	0.71	0.71	0.71
Chronic obstructive pulmonary disease	**Lung infiltrates**	N/A	N/A	N/A
Rheumatoid arthritis	Lupus	0.70	0.71	0.70
Brain tumor	Intracranial hemorrhage	0.69	0.68	0.69
Carpal tunnel syndrome	Osteoarthritis	0.66	0.66	0.66
Diabetes mellitus	Hypertension	0.82	0.81	0.81
Acne	Syringe	0.54	0.54	0.54
Antibiotic	Allergy	0.67	0.67	0.67
Cortisone	Total knee replacement	0.47	0.44	0.47
Pulmonary embolus	**Myocardial infarction**	N/A	N/A	N/A
Pulmonary Fibrosis	Lung Cancer	0.72	0.70	0.72
Cholangiocarcinoma	Colonoscopy	0.63	0.62	0.61
Lymphoid hyperplasia	Laryngeal Cancer	0.70	0.70	0.70
Multiple Sclerosis	Psychosis	0.69	0.67	0.67
Appendicitis	Osteoporosis	0.55	0.55	0.54
Rectal polyp	**Aorta**	N/A	N/A	N/A
Xerostomia	Alcoholic cirrhosis	0.67	0.67	0.66
Peptic ulcer disease	Myopia	0.47	0.47	0.48
Depression	Cellulitis	0.55	0.54	0.54
Varicose vein	**Entire knee meniscus**	N/A	N/A	N/A
Hyperlipidemia	Metastasis	0.56	0.55	0.55

Table 2: Results of Word vector representations.

generated by UMLS-Similarity tool (McInnes et al., 2009).

4   Results and Discussion

Table 2 shows the term pairs in the dataset and the similarity of the terms determined by our measure using three different vector representations. Table

Measures		Physician	Coder
path	Nguyen and Al-Mubaid	0.627	0.852
path	UMLS-Similarity	0.486	0.581
lch	Nguyen and Al-Mubaid	0.672	0.856
lch	UMLS-Similarity	0.486	0.581
wup	Nguyen and Al-Mubaid	0.652	0.794
wup	UMLS-Similarity	0.453	0.535
nam	Nguyen and Al-Mubaid	0.666	**0.862**
nam	UMLS-Similarity	0.448	0.551
Vector representation	Without retrofitting	0.646	0.632
Vector representation	Retrofitted with UMLS-Similarity	**0.696**	0.665
Vector representation	Retrofitted with MeSH tree structure	0.675	0.655

Table 3: Spearman's Rank Correlation Results. Our results are compared with the results of these different measures reported by Nguyen and Al-Mubaid and also generated by UMLS-Similarity tool. path:path based similarity measure; lch: the similarity measure proposed by Leacock & Chodorow in 1998; wup: the similarity measure proposed by Wu & Palmer in 1994; nam:the similarity measure proposed by Nguyen & Al-Mubaid in 2006

3 shows the correlation results between our methods and the judgments made by physicians and coders, as well as the results reported by Nguyen, et al. (Nguyen and Al-Mubaid, 2006) and McInnes, et al. (McInnes et al., 2009), using the UMLS-Similarity tool.

From Table 3, we can see our retrofitted vector representation with UMLS-Similarity obtains a highest correlation with the Physician judgments. Though our retrofitted vector representation achieved a lower correlation with the Coder judgments than the results reported by Nguyen and Al-Mubaid (Nguyen and Al-Mubaid, 2006), we still see an improvement from the retrofitted vector representations as compared with the original vector representation without retrofitting. Since UMLS-Similarity's results are lower than our vector representations, it is understandable that our retrofitted vector representations still can not surpass the results achieved by Nguyen and Al-Mubaid's method. From Table 3, we can also see that our vector representations obtain lower correlations with the coder judgments than with the physician judgments. This contrasts with both the UMLS-Similarity results and those reported by Nguyen and Al-Mubaid. We believe that the reason for this phenomenon is that the coder group were more familiar with the ontology or taxonomy than the physician group. When reviewing these pairs of concepts, coders may interpret the terms in relation to the ontology or taxonomy, whereas physicians may be more likely to understand them at a broader contextual level. Because our vector representation methods all originated as context vectors, this may explain why our methods achieved higher correlation with physician judgments.

Among the three types of vector representations, the retrofitted vector representation with UMLS-Similarity had a higher correlation with both physician and coder judgments than the vectors retrofitted using the MeSH tree structure. We believe this occurred because the way we created the semantic lexicon from the MeSH tree structure had a limited effect on the original vector representations. From Table 2, we can see that the semantic lexicon based on the MeSH tree structure only affected 10 of 25 pairs of terms. The semantic lexicon based on UMLS-Similarity results affected 16 of 25 pairs of terms. We used the MeSH term's parents and children as the lexicon terms, and it is unlikely for a PubMed article to be indexed with both parent and child terms. The UMLS-Similarity approach is more permissive. Two MeSH terms are accepted as a lexicon term only when they have above-threshold similarity as estimated by path-based measures.

5 Conclusions and Future Work

In this paper, we introduced a semantic similarity measure that utilizes vector word representation and the linkage information in an ontology or taxonomy. By retrofitting vector representations with additional ontology or taxonomy information, we can generate vector representations in which lexically-linked concepts are more likely to have similar vector representations. This leads to better approximation of human judgments on the task of estimating semantic relatedness. We show that our method obtains a higher correlation with physician judgments than UMLS-Similarity, and previously reported results. We also demonstrate a clear improvement from the retrofitted vector representation as compared to the vector representation without retrofitting. In the future we plan to expand this technique to other knowledge sources and datasets. We also plan to use more sophisticated and better established approaches to generate concept vectors, e.g. methods of distributional semantic (Cohen et al., 2010), word embedding (Mikolov et al., 2013), and compare with more recently evaluations using neural network based similarity and relatedness measures (Pakhomov et al., 2016).

Acknowledgments

This work was supported in part by the UTHealth Innovation for Cancer Prevention Research Training Program Predoctoral Fellowship (Cancer Prevention and Research Institute of Texas (CPRIT) grant # RP140103), NIH NCATS grant UL1 TR000371, NIH NCI grant U01 CA180964 and the Brown Foundation. The content is solely the responsibility of the authors and does not necessarily represent the official views of the CPRIT.

References

Alan R. Aronson and François-Michel Lang. 2010. An overview of metamap: historical perspective and recent advances. *JAMIA*, 17(3):229–236.

Montserrat Batet, David Sánchez, and Aida Valls. 2011. An ontology-based measure to compute semantic similarity in biomedicine. *J. of Biomedical Informatics*, 44(1):118–125, February.

David M. Blei, Andrew Y. Ng, and Michael I. Jordan. 2003. Latent dirichlet allocation. *J. Mach. Learn. Res.*, 3:993–1022, March.

Olivier Bodenreider and Anita Burgun. 2004. Aligning knowledge sources in the umls: methods, quantitative results, and applications. *Studies in health technology and informatics*, 107(0 1):327.

Jorge E Caviedes and James J Cimino. 2004. Towards the development of a conceptual distance metric for the umls. *Journal of biomedical informatics*, 37(2):77–85.

Massimiliano Ciaramita, Aldo Gangemi, and Esther Ratsch. 2008. Unsupervised learning of semantic relations for molecular biology ontologies. *Ontology Learning and Population: Bridging the Gap between Text and Knowledge*, pages 91–107.

Trevor Cohen, Roger Schvaneveldt, and Dominic Widdows. 2010. Reflective random indexing and indirect inference: A scalable method for discovery of implicit connections. *Journal of biomedical informatics*, 43(2):240–256.

Manaal Faruqui, Jesse Dodge, Sujay K Jauhar, Chris Dyer, Eduard Hovy, and Noah A Smith. 2014a. Retrofitting word vectors to semantic lexicons. *arXiv preprint arXiv:1411.4166*.

Manaal Faruqui, Jesse Dodge, Sujay K Jauhar, Chris Dyer, Eduard Hovy, and Noah A Smith. 2014b. Retrofitting word vectors to semantic lexicons code.

Jay J Jiang and David W Conrath. 1997. Semantic similarity based on corpus statistics and lexical taxonomy. *arXiv preprint cmp-lg/9709008*.

Claudia Leacock and Martin Chodorow. 1998. Combining local context and wordnet similarity for word sense identification. *WordNet: An electronic lexical database*, 49(2):265–283.

Yongjing Lin, Wenyuan Li, Keke Chen, and Ying Liu. 2007. A document clustering and ranking system for exploring medline citations. *Journal of the American Medical Informatics Association*, 14(5):651–661.

Dekang Lin. 1998. An information-theoretic definition of similarity. In *Proceedings of the Fifteenth International Conference on Machine Learning*, ICML '98, pages 296–304, San Francisco, CA, USA. Morgan Kaufmann Publishers Inc.

Andrew L. Maas, Raymond E. Daly, Peter T. Pham, Dan Huang, Andrew Y. Ng, and Christopher Potts. 2011. Learning word vectors for sentiment analysis. In *Proceedings of the 49th Annual Meeting of the Association for Computational Linguistics: Human Language Technologies - Volume 1*, HLT '11, pages 142–150, Stroudsburg, PA, USA. Association for Computational Linguistics.

Bridget T McInnes, Ted Pedersen, and Serguei VS Pakhomov. 2009. Umls-interface and umls-similarity:

open source software for measuring paths and semantic similarity. In *AMIA Annual Symposium Proceedings*, volume 2009, page 431. American Medical Informatics Association.

Bridget T McInnes, Ted Pedersen, Ying Liu, Genevieve B Melton, and Serguei V Pakhomov. 2014. U-path: An undirected path-based measure of semantic similarity. In *AMIA Annual Symposium Proceedings*, volume 2014, page 882. American Medical Informatics Association.

MeSH. 2016. National library of medicine (nlm), mesh browser.

Tomas Mikolov, Ilya Sutskever, Kai Chen, Greg S Corrado, and Jeff Dean. 2013. Distributed representations of words and phrases and their compositionality. In *Advances in neural information processing systems*, pages 3111–3119.

Hoa A Nguyen and Hisham Al-Mubaid. 2006. New ontology-based semantic similarity measure for the biomedical domain. In *Granular Computing, 2006 IEEE International Conference on*, pages 623–628. IEEE.

Serguei Pakhomov, Bridget McInnes, Terrence Adam, Ying Liu, Ted Pedersen, and Genevieve B Melton. 2010. Semantic similarity and relatedness between clinical terms: an experimental study. In *AMIA annual symposium proceedings*, volume 2010, page 572. American Medical Informatics Association.

Serguei VS Pakhomov, Ted Pedersen, Bridget McInnes, Genevieve B Melton, Alexander Ruggieri, and Christopher G Chute. 2011. Towards a framework for developing semantic relatedness reference standards. *Journal of biomedical informatics*, 44(2):251–265.

Serguei VS Pakhomov, Greg Finley, Reed McEwan, Yan Wang, and Genevieve B Melton. 2016. Corpus domain effects on distributional semantic modeling of medical terms. *Bioinformatics*, page btw529.

Siddharth Patwardhan. 2003. *Incorporating dictionary and corpus information into a context vector measure of semantic relatedness*. Ph.D. thesis, University of Minnesota, Duluth.

Siddharth Patwardhan. 2006. Using wordnet-based context vectors to estimate the semantic relatedness of concepts. In *In: Proceedings of the EACL*, pages 1–8.

Ted Pedersen, Siddharth Patwardhan, and Jason Michelizzi. 2004. Wordnet:: Similarity: measuring the relatedness of concepts. In *Demonstration papers at HLT-NAACL 2004*, pages 38–41. Association for Computational Linguistics.

Ted Pedersen, Serguei V. S. Pakhomov, Siddharth Patwardhan, and Christopher G. Chute. 2007. Measures of semantic similarity and relatedness in the biomedical domain. *J. of Biomedical Informatics*, 40(3):288–299, June.

Roy Rada, Hafedh Mili, Ellen Bicknell, and Maria Blettner. 1989. Development and application of a metric on semantic nets. *Systems, Man and Cybernetics, IEEE Transactions on*, 19(1):17–30.

Philip Resnik. 1995. Using information content to evaluate semantic similarity in a taxonomy. In *Proceedings of the 14th International Joint Conference on Artificial Intelligence - Volume 1*, IJCAI'95, pages 448–453, San Francisco, CA, USA. Morgan Kaufmann Publishers Inc.

Peter D. Turney. 2005. Measuring semantic similarity by latent relational analysis. In *Proceedings of the 19th International Joint Conference on Artificial Intelligence*, IJCAI'05, pages 1136–1141, San Francisco, CA, USA. Morgan Kaufmann Publishers Inc.

Zhibiao Wu and Martha Palmer. 1994. Verbs semantics and lexical selection. In *Proceedings of the 32Nd Annual Meeting on Association for Computational Linguistics*, ACL '94, pages 133–138, Stroudsburg, PA, USA. Association for Computational Linguistics.

Zhiguo Yu, Todd R Johnson, and Ramakanth Kavuluru. 2013. Phrase based topic modeling for semantic information processing in biomedicine. In *Machine Learning and Applications (ICMLA), 2013 12th International Conference on*, volume 1, pages 440–445. IEEE.

Zhiguo Yu, Elmer Bernstam, Trevor Cohen, Byron C Wallace, and Todd R Johnson. 2016. Improving the utility of mesh⊛ terms using the topicalmesh representation. *Journal of biomedical informatics*, 61:77–86.

Unsupervised Resolution of Acronyms and Abbreviations in Nursing Notes Using Document-Level Context Models

Katrin Kirchhoff
Department of Electrical Engineering
University of Washington
kk2@u.washington.edu

Anne M. Turner
Department of Health Services
Department of Biomedical Informatics
and Medical Education
University of Washington
amturner@uw.edu

Abstract

Automatic simplification of clinical notes continues to be an important challenge for NLP systems. A frequent obstacle to developing more robust NLP systems for the clinical domain is the lack of annotated training data. This study investigates unsupervised techniques for one key aspect of medical text simplification, viz. the expansion and disambiguation of acronyms and abbreviations. Our approach combines statistical machine translation with document-context neural language models for the disambiguation of multi-sense terms. In addition we investigate the use of mismatched training data and self-training. These techniques are evaluated on nursing progress notes and obtain a disambiguation accuracy of 71.6% without any manual annotation effort.

1 Introduction

As part of a general trend towards patient-centered care many healthcare systems in the U.S. are starting to provide patients with expanded access to clinical notes, often through patient portals connected to their electronic medical record (EMR) systems. Recent studies, such as the OpenNotes project (Delbanco et al., 2012), have found that that patients with access to their health records are more involved in their care and have a better understanding of their treatment plan (Esch et al., 2016; Wolff et al., 2016). However, medical notes often contain complex technical language and medical jargon, requiring patients to seek additional help for linguistic clarification (Walker et al., 2015). Natural language processing (NLP) has the potential to bridge the gap between increased access to medical information and the lack of domain-specific medical training on the patient side. However, in spite of previous work in this area, medical text simplification systems are still not sufficiently mature to be routinely deployed in practice. One problem is the large variety of medical sub-disciplines and document types that need to be covered; another is the lack of annotated training data, often due to constraints on data sharing for reasons of patient privacy.

In this study we investigate *unsupervised* statistical NLP techniques to address one key aspect of medical text simplification, viz. the expansion of medical acronyms and abbreviations (AAs). In addition to text simplification, AA resolution can also help a variety of downstream information extraction tasks. While AA resolution has been studied extensively in the biomedical domain, studies on clinical text are comparatively rare. Moreover, most previous studies use traditional supervised machine learning techniques, consisting of feature extraction and supervised classifiers such as naive Bayes or Support Vector Machines (SVMs) that utilize a carefully developed AA sense inventory and a large amount of hand-annotated ground-truth data. In spite of recently developed methods for rapid data acquisition (crowdsourcing), obtaining reliable manual annotations for highly specialized domains is still difficult and acts as a bottleneck in the development of high-quality medical text simplification systems.

Our proposed approach combines automatic mining of AAs and their possible expansions from medical websites, a first-pass simplification step using

Proceedings of the Seventh International Workshop on Health Text Mining and Information Analysis (LOUHI), pages 52–60,
Austin, TX, November 5, 2016. ©2016 Association for Computational Linguistics

statistical machine translation, and a second-pass rescoring step using recently-developed document-level neural language models. To address the data sparsity issue we investigate model training with mismatched training data as well as self-training.

We evaluate our approach on a subset of a publicly available corpus of nursing progress notes from the MIMIC-II database. Results show an F1 score for AA identification of 0.96, an overall expansion accuracy of 74.3%, and a disambiguation accuracy of 71.6%, all without any supervised annotations used during training.

2 Prior Work

AA identification and resolution has a long history in the biomedical domain. Inventories of AAs and their full forms have been compiled by rule-based (Ao and Takagi, 2005) or machine learning techniques (Movshovitz-Attias and Cohen, 2012; Henriksson et al., 2014; Okazaki et al., 2010), often aided by the fact that biomedical texts tend to define AAs at their first mention. Disambiguation of biomedical AAs has been achieved using traditional machine learning approaches, such as vector space methods (Stevenson et al., 2009), naive Bayes classifiers (Bracewell et al., 2005; Stevenson et al., 2009), and SVMs (Joshi et al., 2006; Stevenson et al., 2009). Clustering has also been used for the purpose of disambiguation (Okazaki and Ananiadou, 2006).

Studies on AAs in clinical text are rarer than those for biomedical texts. In (Pakhomov et al., 2005), disambiguation of clinical AAs was achieved using decision trees and maximum entropy models trained on bag-of-word features from hand-annotated and web-collected text. Moon et al. (2012; 2015) similarly investigated several supervised machine learning techniques and text features for disambiguation of AAs in clinical text, including naive Bayes classifiers, SVMs and decision trees trained on bag-of-word features or Unified Medical Language System (UMLS) concepts. They also noted general problems with AA disambiguation in clinical text, such as shortage of training data due to patient privacy constraints, lack of resources developed for clinical text, and non-standard and highly variable language use in clinical notes. Wu et al. (2015) extended SVM

resp care note : pt on nrb mask + 6l nc required nt sx due inability to clear secretions.

sx copious th yellow sput.

pt sats didn't recover after sx + a&a tx.

Figure 1: Sample nursing note.

classification with vectors based on neural word embeddings. Several systems that participated in the ShaRe/CLEF eHealth Challenge Task on AA normalization (Mowery et al., 2016) utilized conditional random fields (e.g.,(Wu et al., 2013)). Customized expansion dictionaries for clinical text were added in (Xia et al., 2013).

Finally, AA identification and expansion for general English has been addressed by (Ammar et al., 2011; Tevana et al., 2013; Ahmed et al., 2015), among others. The studies most closely related to ours are (Ahmed et al., 2015), which uses language modeling techniques (though not at the document level), and (Ammar et al., 2011), which makes use of statistical machine translation.

3 Data and Task

Our test data consists of nursing progress notes from the MIMIC-II database (Saeed et al., 2011), written by nurses in a cardiac intensive care unit. This data set was chosen because (a) it is publicly available[1]; (b) the documents contain a very high percentage of AAs, thus presenting the problem in a condensed form ; (c) it presents interesting additional challenges: it is characteristic of a highly specialized medical sub-domain, and it contains frequent misspellings, non-standard use of AAs, and elliptical syntax, which we plan to address in future work. The present study is intended to serve as the first step in a more comprehensive simplification system for challenging clinical texts. A sample from a nursing note is shown in Figure 1. AAs are not marked as such – the original documents are either all lowercased, all uppercased, or mixed-case with inconsistent casing; acronyms are not marked by periods. Thus, AAs often overlap in form with regular words, in particular function words – e.g., *is* can be the function word "is" or an abbreviation for *incentive spirometry*.

---

[1]https://mimic.physionet.org/

# words	% AAs	% ambig.
dev set		
125.6 (± 104.3)	25.4 (± 20.4)	73.8 (± 13.6)
eval set		
123.7 (± 112.4)	24.8 (± 9.9)	75.1 (± 12.2)

Table 1: Average number (and stddev) of words, percentage of AAs, and percentage of ambiguous AAs per document, for development (dev) and evaluation (eval) sets.

We use a total of 30 documents (written by various nurses) as reference material. These were split into 15 development and 15 evaluation documents and were manually expanded by medically trained annotators (two medical specialists, one of whom was a hospitalist, and two RNs as additional consultants). The annotation was a consensus annotation; thus, inter-annotator agreement was not measured. The total number of unique AAs in this set is 229, with 611 different instances. Table 1 shows the averages and standard deviations of the number of words, percentage of AAs, and percentage of ambiguous AAs per document. We see fairly large variation in the length of documents and percentages of AAs. On average, however, roughly a quarter of all words are AAs, and 75% of these are ambiguous. We use two other clinical data sets as additional training data: a set of 696 hospital discharge summaries from the i2b2 challenge task (Uzuner et al., 2007) (henceforth "i2b2") and a corpus of 2,365 clinical notes (doctor's notes, hospital discharge summaries, autopsy reports, etc.) from the iDASH repository[2] (henceforth "Cases").

4 Unsupervised Resolution of Abbreviation and Acronyms

Our proposed approach resolves AAs in a largely unsupervised way, requiring true AA sense labels only for system tuning and evaluation but not for training. The first step towards this goal is the acquisition of possible mappings of AAs to their expanded forms. The second step involves preprocessing the nursing documents and generating multiple expanded versions by considering possible combinations of expansions at the sentence level. In a third step, hypotheses are rescored by a document-level language model in order to achieve better dis-

[2] http://dx.doi.org/10.15147/J2H59S

# mappings	9,852
# unique AAs	4,608
# ambiguous AAs	2,817

Table 2: Number of term mappings (total, unique, and ambiguous) extracted from medical terminology websites.

ambiguation and selection of expansions.

4.1 Collecting Term Mappings

The first step towards AA resolution is the collection of a glossary that maps AAs to their expanded forms. We found that existing clinical sense inventories did not provide good coverage for the more specialized domain of ICU nursing – e.g., the clinical sense inventory of (Moon et al., 2012) only covered 7% of the AAs in our development and test data; even the much larger ADAM database of MEDLINE abbreviations (Zhou et al., 2006) covered only 65%. Therefore, we are interested in exploring the feasibility of extracting term mappings automatically from generally accessible resources, without additional human curation. Lists of medical and nursing abbreviations were collected from more than a dozen websites, such as Wikipedia's List of Medical Abbreviations, NIH Medline Plus, ECommunity Health Network, etc., by extracting AAs from html and pdf documents using semi-automated scripts. Note that in order to ensure wide coverage, websites were not restricted to those with nursing terminology; neither was the search biased to maximize coverage of the AAs in our corpus. Rather, we aimed at including a wide range of medical AAs to ensure future reusability for other tasks and domains. A total of 10k mappings were collected; after cleaning and removing duplicates the total number was 9,852. These include medical acronyms and abbreviations, but also health insurance terms, proper names, drug names, etc. The resulting mappings were not hand-curated, annotated, or selected for relevance, in order to minimize the amount of human labor involved. The resulting number of unique AAs is 4,608. 2,817 AAs (61.1%) of these have more than one possible expansion. The maximum number of different expansions is 10; the average is 2.6. As an example, the abbreviation *pt* has the following long forms: *patient, physical therapy, physical therapist, patient teaching, pint, prothrombin time, protime*. Note that we accepted all possible expansions gathered from the websites as valid; we also did not attempt to cluster potential minor variants (like *protime* and *prothrombin time*) into single entries. Although such cleaning steps might improve results, our goal was to evaluate the performance of our approach with potentially noisy data. The final list of term mappings was found to cover 89% of the AAs in our development and test data.

4.2 Term Expansion

The documents are preprocessed by tokenization of punctuation and mapping all numbers to a generic symbol. To create initial expanded versions of our nursing documents with different possible term expansions we utilize a phrase-based statistical machine translation (SMT) system. An SMT system generates target-language translations from source-language input by finding the maximum-likelihood sentence hypothesis obtained by concatenating individual phrase-level translations. The final score for each hypothesis is provided by a log-linear model that computes a weighted sum of feature functions defined on the input s, the output t, or both:

$$score(s, t) = \frac{1}{Z} exp(\sum_k \lambda_k f_k(s, t)) \quad (1)$$

where $f(s, t)$ is a feature function, λ is a weight, and Z is a normalization factor. At a minimum, translation scores and a target-side language model score are included; additional feature functions providing e.g., reordering scores or global coherence scores can be added.

Our system maps 'source' (abbreviated) terms to 'target' (expanded) terms according to a phrase table with all pairs of AAs and their expanded forms, trained from the list of term mapping collected in the first step. No entries are included for AAs that are identical to function words such as *is*, *of*, *on*, etc., as initial development experiments showed that these would lead to an overly high number of false alarms. The drawback is that these AAs will never be expanded and will necessarily count as misses.

The language model in the SMT system is a back-off n-gram model trained using modified Kneser-Ney smoothing. The n-gram order was varied between 3 and 5 and optimized on the development set.

We compared several language models: one trained on the target side of our term mapping list plus i2b2 data, another on the target side plus Cases data, and a third trained on all three.

The maximum phrase length in our translation system is 5. During decoding, no reordering is permitted. The decoding pass generates up to 100 hypotheses per sentence, in order to explore all possible combinations of AA expansions in a sentence.

5 Self-training

Self-training is a general way of utilizing unsupervised data in a classification system. Starting with a system trained on limited data, the system is applied to unlabeled data. The system's predictions are then filtered according to the probability or confidence of the prediction, and the most likely or confident hypotheses are added back to the training data. This procedure can be iterated. Self-training has been used in NLP for e.g., parsing (McClosky et al., 2006) and machine translation (Ueffing et al., 2007). In the context of AA resolution, (Pakhomov, 2002) has used a similar approach to enrich the training data for a maximum entropy classifier.

Here, we use the top-1 hypotheses of our first-pass SMT system to generate additional training data for both the first and second pass language models. To this end we apply the SMT system to the i2b2 and Cases data. Additionally we utilize up to 2000 nursing notes from the MIMIC-II corpus that do not overlap with our development or evaluation sets. One-best hypotheses are generated from our initial SMT system, and are combined with the target side of the term mapping list. This set is then used to retrain the back-off n-gram model used in the SMT system, and to re-generate the first-pass n-best lists. The automatically expanded data is also used to train the document-level language models described in the following section.

6 Document-Level Context Modeling

The selection of appropriate AA expansions is primarily dependent on the the specific medical domain (nursing, cardiology). AA disambiguation could be aided by a detailed sense inventory with domain labels – however, such a classification was not available from our web sources, and considerable manual labor would be required for manual annotation.

As an alternative information source it might be advantageous to take into account not only the local sentence context but also the more global document context. For example, the probability of expanding *hr* to *heart rate* rather than *hour* might be boosted by the occurrence of words such as *cardiovascular* or *blood pressure* earlier in the document. Thus, the document context might serve as a proxy for explicit domain or topic models.

To this end we explore document-context language models (DCLMs) as developed by and described in (Ji et al., 2015). DCLMs are neural language models that attempt to predict words based not only on the local n-gram context as in standard back-off language models, but based on the entire history up to the beginning of the document. Various DCLM architectures have been proposed. We provide a concise summary here; details can be found in (Ji et al., 2015).

General recurrent neural language models (RNNLMs) compute the probability of an output vector (probabilities over the output vocabulary) y at time step n as

$$y_n = softmax(W_h h_n + b) \qquad (2)$$
$$h_n = g(h_{n-1}, x_n) \qquad (3)$$

where W is a weight matrix, b is a bias term, $h \in \mathbb{R}^H$ is a hidden state vector, $x \in \mathbb{R}^K$ is a continuous embedding vector representing the word, and g is a nonlinear activation function. The number of parameters in the network is determined by the dimensionalities of the embedding vector, K, and that of the the hidden vector, H. In "context-to-hidden" DCLMs the hidden state vector in sentence t at time step n is computed not only from the current embedding vector x_n and the preceding state vector $h_{t,n-1}$ but additionally from the last hidden state vector (context vector) of the preceding sentence, $c_{t-1} = h_{t-1,M}$, where M is the last word in the previous sentence. The context vector is simply concatenated with the current embedding vector:

$$h_{t,n} = g(h_{n-1}, x_n \circ c_t) \qquad (4)$$

Alternatively, the context vector can be directly combined with the output vector ("context-to-output" model), using its own weight matrix:

$$y_{t,n} = softmax(W_h h_{t,n} + W_c c_{t-1} + b) \qquad (5)$$

Due to the addition of a second weight matrix W_c this model has more parameters and may be more difficult to train on limited data. Finally, an "attention-based" architecture has been proposed to address the limits of a fixed-dimensional representation of variably-sized document contexts by formulating the context vector as a linear combination of all hidden states in the previous sentence:

$$c_{t-1,n} = \sum_{m=1}^{M} \alpha_{m,n} h_{t-1}, m \qquad (6)$$

Thus, the model can attend to different words in the previous sentence selectively. Moreover, a different context vector is computed for every word n in the current sentence. The context vector is added to both the hidden and the output representation for the current sentence. While this creates a more flexible model, the number of parameters also increases greatly.

Different DCLMs, as well as standard RNNLMs, and RNNLMs whose context can extend beyond the previous sentence boundary, were implemented[3] and were trained using AdaGrad optimization on the same data sets as the back-off ngram models used in the SMT system. 90% of the data was used for training while 10% were held out as development data. Training was stopped when the difference in development set perplexity between the previous and the current iteration was at most 0.5. Different values were investigated for the number units in the embedding and hidden layers (K and H).

For second-pass rescoring of n-best lists with DCLMs we proceed as follows. For each hypothesis in the n-best list for the current sentence, a new "document" is created by concatenating the hypothesis with the previous document context. Each of these documents is scored with the DCLM. The hypothesis resulting in the lowest per-document perplexity chosen and committed to the growing document context. Since no prior context is available for the first sentence in each document, and all further choices are dependent on the choices for previous

---

[3]Using https://github.com/jiyfeng/dclm

sentences, we choose the 1-best hypothesis from the first pass SMT system for the first sentence, rather than assuming a "dummy" context. The vocabulary of the models is restricted to those words that occur at least 3 times in the training data; all others are mapped to a generic "unknown word" symbol.

We noticed during training that the attention-based DCLM obtained much higher perplexity on the development data than the other models, most likely as a result of having too little training data in relation to the number of parameters. This model was therefore excluded from further experiments.

7 Experiments and Results

The first evaluation criterion for our method is the correct identification of AAs vs. regular words. Contrary to rule-based or supervised approaches to AA identification (Nadeau and Turney, 2005; Dannélls, 2006; Moon et al., 2015) AAs are not identified explicitly but implicitly through the choices made by the SMT system. AA identification can be considered a binary detection problem and can thus be evaluated by precision, recall, and F1 score. The second evaluation measure is *overall accuracy*, i.e., the overall percentage of correct AA expansions. Finally, we measure the *disambiguation accuracy*, i.e., the percentage of correct expansions of ambiguous AAs only.

Table 3 shows precision (P), recall (R), F1-score (F1), overall accuracy (A) and disambiguation accuracy (DA) on the eval set for several baseline systems. *Random* is a baseline system where one of the sentence hypotheses produced by the SMT system is selected randomly.[4] Precision and recall are high (and generally stable across all different models), since it is only a small number of words not caught by the function word filter that are consistently misinterpreted as AAs. *Oracle* refers to results obtained by a system that always chooses the hypothesis yielding the highest disambiguation accuracy according to the reference annotation – this represents the upper bound on the accuracy that can be achieved given our automatically collected term mappings. The gap between the oracle accuracy and

---

	System	P	R	F1	A	DA
1	Random	0.95	0.97	0.96	56.6	48.2
2	Oracle	0.93	0.95	0.94	80.0	78.6
3	SMT	0.95	0.97	0.96	72.0	68.0

Table 3: Precision (P), recall (R), F1-score (F1), overall accuracy (A) and disambiguation accuracy (DA) for random baseline, oracle topline, and 1-best output from initial SMT system.

	System	P	R	F1	A	DA
	+ self-training					
1	Random	0.95	0.97	0.96	60.4	52.4
2	Oracle	0.96	0.97	0.96	80.9	80.7
3	SMT	0.95	0.97	0.96	72.2	69.4
	+ DCLM					
4	DCLM	0.95	0.97	0.96	74.3	71.6

Table 4: Precision (P), recall (R), F1-score (F), overall expansion accuracy (A) and disambiguation accuracy (DA) after self-training and second-pass rescoring with DCLMs.

100% accuracy is due to missing expansions in our term mapping list. Row 3 in Table 3 is the result obtained by the first-pass SMT system. The LM for this system was optimized on the development set and consists of a 4-gram back-off model trained using modified Kneser-Ney smoothing on the combined Cases and i2b2 data and the target side of our term mapping list. Accuracy scores obtained by the SMT model are markedly higher than random scores, though there is still much room for improvement.

Table 4 shows the results obtained by an improved system that utilizes self-training and DCLMs. For self-training, the amount of automatically expanded MIMIC-II data and the combination with Cases and i2b2 data was optimized on the development set. Combining the latter two sets with 1,500 expanded documents from MIMIC to train a 4-gram back-off LM was found to be best. Since new n-best lists are generated using the self-trained models, the Random and Oracle results are different (and improved). The accuracy of our SMT system's output is also improved by 1.4% absolute.

For rescoring hypotheses with document-level language models we investigated the DCLM architectures described in Section 6, minus the attention-based model, well as standard RNNLMs and RNNLMs whose context can extend in the past beyond the sentence boundary. The number of parame-

---

[4]A majority sense baseline system is not available due to the lack of a sense inventory with frequency information for this data set.

ters for each model (K and H) was optimized on the development set. Different models trained on different automatically expanded data sources (Cases, i2b2, and MIMIC-II) and their combinations were investigated. It was found that the combined data as well as the Cases and i2b2 data sets in isolation actually resulted in a *worse* performance of the rescored system compared to the first-pass SMT system. While our mismatched data sources did help in training the SMT system, DCLMs, which attempt to model the entire document structure, seem to be very sensitive to mismatched data. By contrast, DCLMs trained on the automatically expanded MIMIC-II data only did achieve an improvement over the first-pass system. The best model (obtained by development set optimization) was a "context-to-hidden" DCLM with a hidden layer size of 48 and a word embedding layer size of 128. The best final overall accuracy on the evaluation set is 74.3%; the disambiguation accuracy is 71.6%. This is fairly close to the topline disambiguation accuracy of 80.2% that can be achieved given our term inventory; however, there is further room for improvement. Of the different document context models tested, all performed in a similar range – e.g., the best models with other architectures ("context-to-output" and RNNLMs without sentence boundary) achieved between 70.2% and 71.1% disambiguation accuracy on the eval set. Furthermore, an RNNLM model with only the current sentence as context achieves 70.5%. Thus, while DCLMs seem to provide slight improvements, our text sample is currently too small to assess statistically significant differences between different architectures or context lengths. Rather, the benefit seems to derive from the neural probability estimation technique used in RNNLM-style models.

Figure 2 shows the automatically expanded version of the sample in Figure 1. While most expansions were acceptable, our term mapping list did not contain a domain-appropriate entry for *a&a*, which was therefore expanded incorrectly to *arthroscopy and arthrotomy* rather than *albuterol and atroven*.

8 Discussion

In this paper we have explored unsupervised and self-supervised resolution of AAs in nursing notes. Contrary to most previous work, which has utilized

respiratory care note: patient on non-rebreather mask and 6l nasal cannula required nasotracheal suction due inability to clear secretions.
suction copious thick yellow sputum .
patient oxygen/blood saturation level didn't recover after suction and arthroscopy and arthrotomy therapy.

Figure 2: Expanded version of nursing note.

supervised classifiers, AA resolution was achieved using web mining to extract term mappings, statistical machine translation, and document-level neural language modeling. With the exception of a small set of hand-annotated documents used to evaluate different models, no ground truth labels were required. Results demonstrated positive effects from self-training and neural language models. Future work will include leveraging additional sources for term mappings, the development of statistical models to improve syntactic readability, and readability experiments with lay human readers.

Acknowledgments

The 'Cases' dataset used in this project was downloaded from iDASH repository (https://idash-data.ucsd.edu) supported by the National Institutes of Health through the NIH Roadmap for Medical Research, Grant U54HL108460. Deidentified clinical records used in this research were provided by the i2b2 National Center for Biomedical Computing funded by U54LM008748 and were originally prepared for the Shared Tasks for Challenges in NLP for Clinical Data organized by Dr. Ozlem Uzuner, i2b2 and SUNY. This work was supported in part by the National Library of Medicine (NLM) under award number R01 10432704 and by the UW Provosts Office through a grant to the first author.

References

A.G. Ahmed, F.F.A. Hady, E. Nabil, and A. Badr. 2015. A language modeling approach for acronym expansion disambiguation. In *Computational Linguistics and Intelligent Text Processing. Lecture Notes in Computer Science vol. 9041*, pages 264–278.

W. Ammar, K. Darwish, A. El Kahki, and K. Hafez. 2011. ICE-TEA: in-context expansion and translation of English abbreviations. In *Proceedings of CICLing*, pages 41–54.

H. Ao and T. Takagi. 2005. ALICE: an algorithm to extract abbreviations from MEDLINE. *JAMIA 12(5)*, pages 576–586.

D.B. Bracewell, S. Russell, and A.S. Wu. 2005. Identification, expansion and disambiguation of acronyms in biomedical text. In *Proceedings of ISPA Workshops*, pages 186–195.

D. Dannélls. 2006. Automatic acronym recognition. In *Proceedings of EACL*, pages 167–170.

T. Delbanco, J. Walker, S.K. Bell, J.D. Darer, J.G. Elmore, N. Faraq, et al. 2012. Inviting patients to read their doctors notes: Quasi-experimental study and a look ahead. *Annals of Internal Medicine*, 15(7):461.

T. Esch, R. Mejilla, M. Anselmo, B. Podtschaske, T. Delbanco, and J. Walker. 2016. Engaging patients through open notes: an evaluation using mixed methods. *BMJ Open*, 6:e010034.

A. Henriksson, H. Moen, M. Skeppstedt, V. Daudaravicius, and M. Duneld. 2014. Synonym extraction and abbreviation expansion with ensembles of semantic spaces. *Journal of Biomedical Semantics 5(6)*.

Y. Ji, T. Cohn, L. Kong, C. Dyer, and J. Eisenstein. 2015. Document context language models. *CoRR*, abs/1511.03962.

M. Joshi, T. Pedersen, R.Maclin, and S. Pakhomonov. 2006. Kernel methods for word sense disambiguation and acronym expansion. In *Proceedings of AAAI*.

D. McClosky, E. Charniak, and M. Johnson. 2006. Effective self-training for parsing. In *Proceedings of HLT-NAACL*, pages 152–159.

S. Moon, S. Pakhomov, and G. Melton. 2012. Automated disambiguation of acronyms and abbreviations in clinical texts: Window and training size considerations. In *Proceedings of AMIA*, pages 1310–1319.

S. Moon, B. McInnes, and G.B. Melton. 2015. Challenges and practical approaches with word sense disambiguation of acronyms and abbreviations in the clinical domain. *Healthc Inform Res*, 21(1):35–42.

D. Movshovitz-Attias and W. Cohen. 2012. Alignment-HMM-based extraction of abbreviations from biomedical text. In *Proceedings of the BioNLP*, pages 47–55.

D.L. Mowery, B.R. South, L. Christensen, and J. Leng et al. 2016. Normalizing acronyms and abbreviations to aid patient understanding of clinical texts: ShaRe/CLEF ehealth challenge 2013, task 2. *J Biomed Semantics*, 7(43).

D. Nadeau and P. Turney. 2005. A supervised learning approach to acronym identification. In *Canadian Society Conference on Advances in Artificial Intelligence*, pages 319–329.

N. Okazaki and S. Ananiadou. 2006. Clustering acronyms in biomedical text for disambiguation. In *Proceedings of LREC*, pages 959–962.

N. Okazaki, S. Ananiadou, and J. Tsuji. 2010. Building a high-quality sense inventory for improved abbreviation disambiguation. *Bioinformatics*, 26(9):1246–1253.

S. Pakhomov, T. Pedersen, and C.G. Chute. 2005. Abbreviation and acronym disambiguation in clinical discourse. pages 589–593.

S. Pakhomov. 2002. Semi-supervised maximum entropy based approach to acronym and abbreviation normalization in medical texts. In *Proceedings of ACL*, pages 160–167.

M. Saeed, M. Villaroel, A.T. Reisner, and G. Clifford et al. 2011. Multiparameter intelligent monitoring in intensive care II (MIMIC-II): A public access ICU database. *Critical Care Medicine*, 39(5).

M. Stevenson, Y. Guo, A. Al Amri, and R. Gaizauskas. 2009. Disambiguation of biomedical abbreviations. In *Proceedings of the Workshop on BioNLP*, pages 71–79.

B. Tevana, T. Cheng, K. Chakrabarti, and Y. He. 2013. Mining acronym expansions and their meanings using query click log. In *Proceedings of WWW*, pages 1261–1271.

N. Ueffing, G. Haffari, and A. Sarkar. 2007. Transductive learning for statistical machine translation. In *Proceedings of ACL*.

O. Uzuner, U. Juo, and P. Szolovits. 2007. Evaluating the state-of-the-art in automatic de-identification. *JAMIA 14(5)*, pages 550–63.

J. Walker, M. Meltsner, and T. Delbanco. 2015. US experience with doctors and patients sharing clinical notes. *BMJ*, page 350:g7785.

J.L. Wolff, J.D. Darer, A. Berger, and D. Clarke et al. 2016. Inviting patients and care partners to read doctors' notes: OpenNotes and shared access to electronic medical records. *JAMIA*, to appear.

Y. Wu, B. Tang, M. Jiang, S. Moon, J.C. Denny, and H. Xu. 2013. Clinical acronym/abbreviation normalization using a hybrid approach. In *Proceedings of CLEF*.

Y. Wu, J. Xu, Y. Zhang, and H. Xu. 2015. Clinical abbreviation disambiguation using neural word embeddings. In *Proceedings of the 2015 Workshop on Biomedical Natural Language Processing (BioNLP)*, pages 171–176.

Y. Xia, X. Zhong, P. Liu, C. Tan, S. Na, Q. Hu, et al. 2013. Normalization of abbreviations/acronyms: THCIB at CLEF eHealth 2013 Task 2. In *CLEF 2013 Evaluation Labs and Workshops: Working Notes*.

W. Zhou, V.I.Torvik, and N.R. Smalheiser. 2006. ADAM: another database of abbreviations in MEDLINE. *Bioinformatics*, 22(22)::2813–2818.

Low-resource OCR error detection and correction in French Clinical Texts

Eva D'hondt
LIMSI, CNRS
Université Paris-Saclay
F-91405 Orsay
eva.dhondt@limsi.fr

Cyril Grouin
LIMSI, CNRS
Université Paris-Saclay
F-91405 Orsay
cyril.grouin@limsi.fr

Brigitte Grau
LIMSI, CNRS, ENSIIE
Université Paris-Saclay
F-91405 Orsay
bg@limsi.fr

Abstract

In this paper we present a simple yet effective approach to automatic OCR error detection and correction on a corpus of French clinical reports of variable OCR quality within the domain of foetopathology. While traditional OCR error detection and correction systems rely heavily on external information such as domain-specific lexicons, OCR process information or manually corrected training material, these are not always available given the constraints placed on using medical corpora. We therefore propose a novel method that only needs a representative corpus of acceptable OCR quality in order to train models. Our method uses recurrent neural networks (RNNs) to model sequential information on character level for a given medical text corpus. By inserting noise during the training process we can simultaneously learn the underlying (character-level) language model and as well as learning to detect and eliminate random noise from the textual input. The resulting models are robust to the variability of OCR quality but do not require additional, external information such as lexicons. We compare two different ways of injecting noise into the training process and evaluate our models on a manually corrected data set. We find that the best performing system achieves a 73% accuracy.

1 Introduction

While most of the contemporary medical documents are created in electronic form, many of the older patient files are kept in paper version only. These files represent an invaluable source of information and experience for medical investigations, especially in domains with low-frequency diseases such as foetopathology, the medical domain which specializes in the treatment and diagnosis of illnesses in unborn children. Over the last two decades, Optical Character Recognition (OCR) technology has improved substantially which has allowed for a massive institutional digitization of textual resources such as books, newspaper articles, ancient handwritten documents, etc. (Romero et al., 2011).

In recent years, hospitals and medical centers have taken to processing older, paper-based resources into digital form in order to construct knowledge bases and resources that can be consulted by medical staff and students. When it comes to documents containing patient information, however, the process of digitization or the use of the resulting text corpus are not as straightforward as they may seem on first sight. Firstly, medical corpora are much less accessible than other general-purpose text corpora since the confidentiality of patients is a first priority. This results in limited access of researchers to original files which in turns directly limits the quantity of files that can be digitized. Secondly, text corpora that contain medical information can only be distributed (even internally in hospitals or research centers) when they are de-identified, that is, when all patient-specific information is identified and removed from the OCRed text (Richards, 2009). This additional processing step can have a significant impact on the quality of the resulting text corpus when information is incorrectly identified as patient-specific information and consequently trans-

Proceedings of the Seventh International Workshop on Health Text Mining and Information Analysis (LOUHI), pages 61–68,
Austin, TX, November 5, 2016. ©2016 Association for Computational Linguistics

formed or removed, e.g. 'Parkinson' in the phrase 'Parkinson's disease'. A side-effect of the obligation of de-identification is that OCR process information is often not available to the researcher using the text corpus afterwards, since it could potentially be used to reconstruct the original information in the paper version. Thirdly, medical files in hospitals are generated over many years. Consequently, the variations in paper, printing techniques or differences in structuring the text (e.g., one-column versus two-column paper formats) can impact the OCR process, and the quality of OCRed files can vary substantially from one year to another (Evershed and Fitch, 2014).

With the increased use of OCR to digitize paper corpora, the problem of OCR error detection and correction has received considerable attention from the research community, especially as regards to its impact on information retrieval and information extraction tasks (Ruch et al., 2002; Magdy and Darwish, 2010). The majority of the current OCR error correction systems use the same three-step approach: (1) OCR error detection; (2) candidate generation; (3) candidate ranking. In the first step, a potential OCR error is detected using either a lookup in a domain-specific lexicon (Kissos and Dershowitz, 2016) or unigram language model (Bassil and Alwani, 2012), and/or by consulting information from the OCR process, i.e., the confidence scores of the recognized characters. The second step, candidate generation, also heavily depends on external resources, either by generating potential candidate replacements for the erroneous words from a lexicon (Piasecki and Godlewski, 2006) or by learning and using a mapping of characters that were often interchanged during the OCR process to generate potential candidates with string distance metrics (Kukich, 1992). Such mappings are known as 'character confusions' but need to be learned over a training corpus of a considerable size before they can become effective (Evershed and Fitch, 2014). The lack of external information such as OCR process information or domain (and hospital)-specific lexicons and the high variability of OCR quality render these systems useless for OCR error detection in medical text corpora.

Unlike the current state-of-the-art systems, the method proposed in this article requires only a sample of (relatively) clean domain-specific text, and no other external information. It uses recurrent neural networks (RNNs) to train character-level language models. By artificially inserting noise into the training data, the system learns to filter out random noise, while learning the domain-specific language model that underlies the documents in the corpus. Since the models do not depend on external resources the method can also be applied to domain-specific text corpora outside the medical domain, on the condition that the documents in the training corpus are not too heterogeneous.

2   Background

OCR and orthography error detection and correction have received interest from the NLP community since the seventies. A good survey of the early work on this problem can be found in Kukich (1992). While most of the traditional OCR error detection systems focused on the construction of so-called 'confusion matrices' of character (pairs) to detect corruptions of existing words into non-words, more recent systems find that using information on the language context in which the error appears improves accuracy (Evershed and Fitch, 2014). A good example of the latter is the system proposed by Bassil et al. (2012) who use extensive n-gram word and 2-character models from the Google Web 1T 5-Gram data set to identify OCR errors and generate and identify the most plausible replacements. Kissos et al. (2016) studied the relative impact of different information sources by combining features from language models constructed over the training corpus, OCR process information and document context information. They found that bigrams, i.e., localized context information was the most useful feature in OCR correction.

To the best of our knowledge, the only existing OCR error detection and correction systems for medical texts focus on either OCR correction for historical text with adapted language models (Thompson et al., 2015) or OCR recognition of handwritten notes by doctors, which is not surprising given the absence of large OCRed text corpora in this domain. Notable work in this area was carried out by Piasecki et al. (2006) who examined the construction of word-level language models to improve OCR correction of Polish handwritten medical notes. They found that the repetitive character sequences and recurrent structure of medical notes greatly aided the construction of language models but that this positive effect is domain-specific and does not carry over the similar corpora in a different medical subdomain. Like the more generic OCR error detection and correction systems, they also depend on external resources, in this case, an extensive domain-specific lexicon for the detection of errors and generation of candidates.

'Automatic misspelling detection and correction', a subtask related to OCR error detection and correction, has received a lot of attention over the last few years with the increased use of Electronic Health Records (EHRs) in the medical domain. While these tasks have a similar goal, the underlying assumptions are quite different: Character confusions in misspellings are often regular, either due to phonetic misspelling, or due to the proximity of certain letters on a keyboard. OCR errors, however, are often more random and can occur more frequently (Kumar, 2016). Notable work in this domain include Lai et al. (2015) who combine a noisy channel spelling correction approach with an extensive domain-specific dictionary to generate probable misspelling-correction pairs, and Mykowiecka et al. (2006) who use bigram language models to estimate the probability of a misspelling in a given word.

3   Corpus

3.1   Corpus construction

We train and evaluate our system on a data set of French patient notes from the domain of foetopathology. This corpus was assembled and digitized within the context of the Accordys project, and spans a total of 22 years.[1] In total, the corpus contains the files from 2476 individual patients which amounts to 16,573 paper documents. The files were processed with a custom-trained commercial OCR engine, and later de-identified with an in-house de-identification tool (Grouin and Zweigenbaum, 2013). All identifying data were replaced by generic tags with a numerical identifier for all occurrences of the same information in order to maintain the original distribution of tokens along the corpus

(e.g., the tag "DATE-8734" was used for all occurrences of "May 21st, 2016"). There is a substantial amount of redundancy in the corpus: For some documents, several (nearly-identical[2]) copies were added to the patient's folder. It should be noted that the patient notes in the corpus are very similar with regards to their contents: the vast majority of the patient files are either reports of the pathologic examination of fetus and placenta or results of genetic tests. While the style and structure of these reports change over time in the corpus, their content— and consequently much of the terminology used— remain stable.

3.2   OCR quality in corpus

Since the model of the OCR engine which was used to convert the entire corpus was trained on a subset of documents of more recent years (implying good paper quality, clear font, no ink problem, etc.), the OCR quality of the OCRed documents decreases substantially for the older documents. In a test set of 100 randomly selected documents from the corpus, we found that 16.4% of the words[3] did not appear in the Unified Medical Lexicon for French (Zweigenbaum et al., 2005), a word list with specific technical terms. Of these 16.4%, 3.8% pertained to words that were domain-specific terms that has been correctly identified in the OCR process but which did not feature in the UMLF, and 10.8% were words which contained at least one OCR error. The remaining out-of-vocabulary[4] (OOV) words were not classifiable. Table 1 shows a representative example of an OCRed document of mediocre quality in the corpus.

3.3   Training set

For the purposes of training the neural network described in section 4, we needed to provide the model with relatively clean data to learn a reliable language model. We used the proportion of OOV words with regards to the number of words in the document as

---

[1]The files range from 1983 to 2005.

[2]While the original paper documents might be identical, the process of OCR and de-identification has introduced enough noise that very few identical files remain.

[3]We performed simple whitespace tokenization with removal of punctuation to obtain the set of words.

[4]The vocabulary was made up of Unified Medical Lexicon for French and a list of domain-specific terms extracted from a comprehensive French handbook of foetopathology (Bouvier et al., 2008)

I. EXAMEN MACROSCOPIQUE
- fœtus de sexe masculin
- état frais
- macération absente
- poids 440 gr
- **menurations** VT 2/ cm
VC 19 cm PC 19 cm Pied 3,5
- ces paramètres sont compatibles avec un
âge gestationnel de **21'22S,A**

...

La dissection des viscères met en évidence :
- **hvpoplasie** du cœur gauche

...

Les clichés ne montrent pas **danomalie**
osseuse autre que **faOsence Oe** la 12ème
paire de côtes.

Table 1: Feto-placental report sample with fake data and realistic digitization errors. Incorrectly digitized tokens are in bold.

a simple heuristic to determine the OCR quality of the document. Using this metric we divided the corpus into four categories, as shown in Table 2. The right column shows the cut-off rates that were used to distinguish between the different categories. The lower the document score, the fewer OOV words were found which indicates a good OCR quality. We would like to stress that although we use external resources to classify the training corpus into categories, this information is not used during the training of the neural networks.

OCR quality	# of documents	score cut-off
Excellent	1,088 (6.6%)	x <= 0.1
Good	7,694 (46.4%)	0.1 > x <= 0.25
Mediocre	3,595 (21.7%)	0.25 > x <= 0.50
Unusable	4,196 (25.3%)	x > 0.50

Table 2: Distribution of OCR quality categories in the training corpus

3.4 Evaluation set

All evaluations in this paper were carried out on a set of 53 files, randomly selected from the Excellent and Good quality subsets, which had been annotated manually by one annotator in two passes. These annotations were later verified by a second annotator.[5]

---

[5]The role of the second annotator was to check that the existing annotations were correct and consistent. Ergo the annota-

In total, the evaluation contains 473 errors. Table 3 shows the distribution of the four main types of OCR errors in the evaluation set. For each error the annotator provided a corrected string. Consequently, for each document in the evaluation set we had an original version with OCR errors, and a corrected version as the Gold Standard.

error type	#	OCR error ex.	Gold Standard
insertion	38	nuquer	nuque
deletion	69	maroscopique	macro-scopique
substitu-tion	349	extrei/iities	extremities
other	17	e};,.ez ,J2	e};,.ez ,J2[7]

Table 3: Distribution of OCR error types in the evaluation set

4 Model

4.1 Character-based LSTM model

Our model consists of a many-to-many character sequence learning network using Long Short Term Memory nodes (LSTMs). The main idea is that the input sequence, in this case a string of characters, is mapped to a vector which is fed into a recurrent neural network (RNN) to generate the output sequence conditioned on the encoding vector. We use LSTMs[8] (Graves, 2013) as the basic RNN unit since this has shown improved performance on various NLP tasks such as text generation. In our model, we stack two LSTM layers on top of each other: the first level is an encoder that reads the source character sequence and the other is a decoder that functions as a language model and generates the output. We also added a drop-out layer since this has been shown to improve performance (Srivastava et al., 2014). The model was implemented in Keras (Chollet, 2015), a python library for deep learning. Figure 1 shows the network hierarchy.

---

tions were not done independently.

[6]Since the annotators did not have access to the original PDF files to check the original text, it was not possible to generate corrected text for some badly corrupted strings.

[7]Since the annotators did not have access to the original PDF files to check the original text, it was not possible to generate corrected text for some badly corrupted strings.

[8]An excellent low-level introduction to RNNs and LSTMs can be found at http://karpathy.github.io/2015/05/21/rnn-effectiveness/.

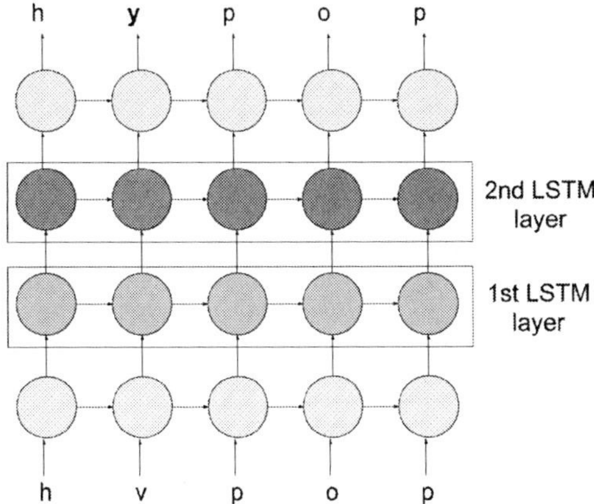

Figure 1: Hierarchy of 2-layer many-to-many sequence learning network; 'hvpop' taken as input, 'hypop' as expected output

In order to learn a robust language model, we fed the neural network with randomly corrupted input strings and provided the original (non-corrupted) strings as output labels. This way the NN learns both a character level language model that is domain-specific but it also learns to detect and eliminate random noise. We created corrupted strings by deleting, inserting and substituting one or two characters for a given string. Since a string could be submitted to multiple corrupting edits this resulted in both mono-error as well as multi-error words in the corrupted string. We heuristically determined the rate of noise so as to resemble the level of corruption, i.e., number of OCR errors of the actual test data. Table 4 shows an example of the generated training input with label output. We used windows of 20 characters from the initial text but since the length of the corrupted text strings varied due insertions and deletions, the network was fed (padded) sequences of 23 characters. The network was trained on data from the 'Excellent' OCR quality subset.

original text (reference)	'après l'expulsion de'
corrupted text (input)	'arpèS1'exVlsion e'
model output	'après 1'exulsion de'

Table 4: Example of input, output and reference in the training process

We experimented with two different string corruption settings:

1. *Random generation (randomNoise)* in which we used a random number generator to determine if and which edit options were selected. Character substitutions were performed at random with characters from the character set;

2. *Insertion of character confusions (confusionPair)*: In this setting we want to examine if injecting information on possible character confusions in the corpus, i.e. teaching the model that character x is likely to be replaced with character y, leads to a faster convergence of the trained models. While we do not have annotated training material to learn character confusions, we can exploit the natural redundancy in the corpus: Using a string alignment algorithm we identified near-duplicates in the subset of documents with 'Good' OCR quality. We then extracted confusion pairs, i.e. 1:1, 2:2, 1:2 and 2:1 character pairs that occurred in the same contexts, and had a relatively high frequency in the corpus. Table 5 shows the top 5 of the most frequent confusion pairs extracted from the corpus. This information was added to the randomization module so that instead of a substitution of a character by a random character, the only substitutions allowed were chosen from this list. We should stress that, since we do not use annotated training material, the extracted list might not be complete.

string to be replaced	replacement string	character pair type
l	I	1:1
I	l	1:1
!	l	1:1
W	VT	1:2
rr	m	2:1

Table 5: Most frequent character confusions from the subset of the corpus with 'Good' OCR quality

4.2 Baseline model

In order to evaluate the relative improvement of our character-based model, we also implemented a traditional word-based OCR error detection and correction system. Our implementation follows the basic structure of such systems which were presented in section 1. The algorithm consists of the following steps:

1. Tokenization of the text into token sequences;

2. OCR error detection by vocabulary[9] look-up. We allowed up to a minimal edit distance of three[10] transformations of a given token, and the combination of the given token glued to the subsequent token in the token sequence[11] to find a suitable entry in the lexicon.

3. The candidates were then ranked and the highest-ranking candidate was used the replace the original token in the text. We experimented with different weighting schemes and finally opted for a ranking by the number of edits, in which substitution edits that used the confusion pairs (presented in section 4.1) had a lower weight than edits which were not significantly present in the training corpus.

5 Experiments

The character-based models were trained for 4 iterations with 20 epochs[12] per iteration. The randomNoise and confusionPair models achieved 73% and 71% accuracy respectively while the baseline model achieved an accuracy of 51%. Inspection of the intermediate scores shows that the randomNoise model achieves convergence fairly quickly, while the confusionPair model has a slower learning rate. This indicates that corrupting the strings in a more 'consistent' manner, i.e. using information on likely confusion pairs extracted from the corpus, leads to more erroneous assumptions during training. While the randomNoise model is trained to robustly deal with random noise, the confusionPair model's focus on a subset of the possible errors does not train the model well enough to detect other kinds of errors.

A close analysis of the corrections and errors of the randomNoise model on the test set shows that the model is good at detecting 'close' substitutions of characters when they appear in a relatively clean environment, e.g. a substitution between 'e' and 'é' in the string 'theorique', or a switch between lowercase and uppercase, such as in 'develoPpement' . We find that when the original input string contains multiple OCR errors close together (and as such is no longer a 'clean' environment for a character substitution), the model cannot adequately decide which characters to replace. This suggests that either gradually increasing the ratio of noise or slowly extending the context window during training might have a positive impact on performance accuracy. Table 6 shows the proportions of OCR errors in the manually annotated evaluation set that were corrected by the two character-based approaches and the word-based baseline model.

OCR error type	randomNoise	confusionPair
insertion	0.0	0.1
deletion	24.5	23.6
substitution	75.5	76.3

Table 6: Proportions of corrections for different OCR error types in the evaluation set

We see that most substitution errors are most easily spotted by the models but that the detection of insertion errors proves very difficult. This is because most of the insertion errors are random insertions of whitespace in words. Since whitespace is used abundantly in structuring the documents, the model generally predicts this character with a high probability, and thus fails to detect it as an error. The addition of character confusion information in the creation of corrupted input data (column 2 in Table 6) has a slight positive impact on substitution errors but not as much as was expected.

When examining the cases in which the model failed to spot an error or generated corrections where none were needed, we find that text written in uppercase presents a great difficulty for the models. Only a small part of the documents are written in uppercase, i.e. the headers with de-identified personal information and the titles of the individual sections. The models clearly do not have enough training data to learn an adequate language model. In a follow-up study, we should either provide the model with more data, or add a lowercasing step to the preprocessing pipeline. Another interesting but infrequent

---

[9]Checks were performed using the same lexicon as for the calculation of the proportion of OOV words in section 3.3. We extended the lexicon by creating new entries which consisted of two original words of the lexicon glued together, in order to catch whitespace deletion errors.

[10]In our implementation insertion, deletion and substitution steps all had the same cost, i.e. 1.

[11]In order to find whitespace insertion errors

[12]The number of epochs was empirically determined.

error are the cases where the language model has clearly learned the character-based language models but uses it incorrectly given the wider context, for example, by changing 'facile' *(easy)* into 'faciale' *(facial)* in 'Ponction de trophoblaste facile' *(easy puncture of the trophoblasts)*. These types of error could be avoided by fitting a larger language model on top of the character-based LSTM model.

6 Conclusion

In this paper we presented a method for the detection and correction of OCR errors in French patient files. Our method consists of a many-to-many sequence learner using LSTMs which is robustly trained on artificially corrupted good-quality training data in order to learn both the underlying character level language model, as well as to detect and eliminate noise in the input string. The relatively fast convergence of the models is likely due to the natural redundancy in the medical corpus. We experimented with two different methods of adding noise to the input and found that injecting information on likely character confusion pairs extracted from the training corpus had no positive impact on accuracy. Interestingly, the models are not good at detecting insertion errors, i.e. the detection of word boundaries. In future work, we would like to extend the model by combining the output of the character level with information on word level through an embedding layer in order to improve the overall accuracy.

Acknowledgments

This work was supported by the French National Agency for Research under the grant Accordys[13] ANR-12-CORD-0007.

References

Youssef Bassil and Mohammad Alwani. 2012. OCR context-sensitive error correction based on google web 1t 5-gram data set. *American Journal of Scientific Research.*

Raymonde Bouvier, Dominique Carles, Marie-Christine Dauge, Pierre Déchelotte, Anne-Lise Delézoide, Bernard Foliguet, Dominique Gaillard, Bernard Gasser, Marie Gonzalès, and Férechté Encha-Razavi. 2008. *Pathologie fœtale et placentaire pratique.* Sauramps Médical, Montpellier.

François Chollet. 2015. Keras. `https://github.com/fchollet/keras`.

John Evershed and Kent Fitch. 2014. Correcting noisy OCR: Context beats confusion. In *Proceedings of the First International Conference on Digital Access to Textual Cultural Heritage*, pages 45–51. ACM.

Alex Graves. 2013. Generating sequences with recurrent neural networks. *arXiv preprint arXiv:1308.0850.*

Cyril Grouin and Pierre Zweigenbaum. 2013. Automatic de-identification of French clinical records: Comparison of rule-based and machine-learning approches. In *Stud Health Technol Inform*, volume 192, pages 476–80, Copenhagen, Denmark. MEDINFO.

Ido Kissos and Nachum Dershowitz. 2016. OCR error correction using character correction and feature-based word classification. In *Proceedings of the 12th IAPR International Workshop on Document Analysis Systems (DAS2016)*, Santorini, Greece.

Karen Kukich. 1992. Techniques for automatically correcting words in text. *ACM Computing Surveys (CSUR)*, 24(4):377–439.

Atul Kumar. 2016. A survey on various OCR errors. *International Journal of Computer Applications*, 143(4):8–10.

Kenneth H Lai, Maxim Topaz, Foster R Goss, and Li Zhou. 2015. Automated misspelling detection and correction in clinical free-text records. *Journal of biomedical informatics*, 55:188–195.

Walid Magdy and Kareem Darwish. 2010. Omni font OCR error correction with effect on retrieval. In *Proceedings of the 10th International Conference on Intelligent Systems Design and Applications*, pages 415–420. IEEE.

Agnieszka Mykowiecka and Małgorzata Marciniak. 2006. Domain–driven automatic spelling correction for mammography reports. In *Proceedings of the Intelligent Information Processing and Web Mining*, pages 521–530, Ustrón, Poland.

Maciej Piasecki and Grzegorz Godlewski. 2006. Language modelling for the needs of OCR of medical texts. In *Proceedings of International Symposium on Biological and Medical Data Analysis*, pages 273–284, Thessaloniki, Greece.

Margaret M Richards. 2009. Electronic medical records: Confidentiality issues in the time of HIPAA. *Professional Psychology: Research and Practice*, 40(6):550.

---

[13] Agrégation de Contenus et de COnnaissances pour Raisonner à partir de cas dans la DYSmorphologie foetale

Verónica Romero, Nicolás Serrano, Alejandro H. Toselli, Joan Andreu Sánchez, and Enrique Vidal. 2011. Handwritten text recognition for historical documents. In *Proceedings of the Language Technologies for Digital Humanities and Cultural Heritage Workshop*, pages 90–96, Hissar, Bulgaria.

Patrick Ruch, Robert Baud, and Antoine Geissbhler. 2002. Evaluating and reducing the effect of data corruption when applying bag of words approaches to medical records. *International Journal of Medical Informatics*, 67(13):75 – 83.

Nitish Srivastava, Geoffrey E Hinton, Alex Krizhevsky, Ilya Sutskever, and Ruslan Salakhutdinov. 2014. Dropout: a simple way to prevent neural networks from overfitting. *Journal of Machine Learning Research*, 15(1):1929–1958.

Paul Thompson, John McNaught, and Sophia Ananiadou. 2015. Customised ocr correction for historical medical text. In *2015 Digital Heritage*, volume 1, pages 35–42. IEEE.

Pierre Zweigenbaum, Robert Baud, Anita Burgun, Fiammetta Namer, Éric Jarrousse, Natalia Grabar, Patrick Ruch, Franck Le Duff, Jean-François Forget, Magaly Douyere, et al. 2005. UMLF: a unified medical lexicon for French. *International Journal of Medical Informatics*, 74(2):119–124.

Citation Analysis with Neural Attention Models

Tsendsuren Munkhdalai, John Lalor and **Hong Yu**
University of Massachusetts, MA, USA
tsendsuren.munkhdalai@umassmed.edu, lalor@cs.umass.edu, hong.yu@umassmed.edu

Abstract

Automated citation analysis (ACA) can be important for many applications including author ranking and literature based information retrieval, extraction, summarization and question answering. In this study, we developed a new compositional attention network (CAN) model to integrate local and global attention representations with a hierarchical attention mechanism. Training on a new benchmark corpus we built, our evaluation shows that the CAN model performs consistently well on both citation classification and sentiment analysis tasks.

1 Introduction

Citations are relations between the cited and citing articles and are important content in literature. There are different reasons that authors choose to cite an article. Identifying the purpose of the citations has important applications including faceted navigation, citation based information retrieval, impact factor assessment and summarization of scientific papers (Hearst and Stoica, 2009).

ACA refers to the tasks of citation function classification and citation sentiment analysis. Pioneered by Garfield and others (1965), a large body of citation-related studies have been carried out to develop categorization schemes for citation function analysis. However, most of the studies are limited to to specific domain. The classification schemes are typically complex, containing multiple overlapping categories ranging from three to 35 (Bornmann and Daniel, 2008). In contrast, the success of ACA

depends on a small but well-defined set of citation categories. Nanba and Okumura (1999) developed a semi-ACA based on a 3-category scheme derived from Garfield and others (1965)'s 15 categories. Similarly, Pham and Hoffmann (2003) developed rule-based approaches (cue phrases) to classify citations into one of the four classes (*basis*, *support*, *limitation* and *comparison*). Teufel et al. (2009) addressed citation function classification and sentiment analysis jointly by a hierarchical scheme with the top nodes for sentiment and the leaf nodes for function classes. Agarwal et al. (2010) developed a scheme of eight non-overlapping categories for citation function classification in biomedical literatures. This scheme simplifies Yu et al. (2009)'s hierarchical overlapping categories. Recently, a decision-tree based scheme was introduced to facilitate citation context based intelligent systems (Mandya, 2012). The citation function classes, organic and perfunctory proposed by Moravcsik and Murugesan (1975) was adapted for a facet-based classification scheme (Jochim and Schütze, 2012).

Machine learning (ML) approaches to ACA mainly adapted statistical classifiers including support vector machines (SVM), logistic regression and Nave-Bayes classifier (Athar, 2011; Athar and Teufel, 2012; Sula and Miller, 2014). The feature set extracted includes n-grams, part-of-speech tags, word stems, cue phrases, sentence dependency components, named entity mentions and word and sentence location based features. Despite the rich linguistically motivated feature sets, ACA remains a challenge, performing significantly worse than human. One of the reasons for this could be the lack of

69

Proceedings of the Seventh International Workshop on Health Text Mining and Information Analysis (LOUHI), pages 69–77,
Austin, TX, November 5, 2016. ©2016 Association for Computational Linguistics

Category	Description
Function Classification	
Background	Citations that describe background of the main topic on the whole, or provide recent studies and state-of-the-art approaches in a general way
Method	Citations of tools, methods, data and other resources used or adapted in the citing work
Results/findings	Citations that authors used to reference others study to relate their research results and/or findings to the cited work
Don't know	This category should be chosen if you dont know which one to select
Sentiment Classification	
Negational	Citations that discuss or dispute the correctness and/or weakness of the cited work
Confirmative	Citations that imply to confirm, support or make use of outcomes of the cited work
Neutral	Citations that are not negational nor confirmative
Don't know	This category should be chosen if you dont know which one to select

Table 1: Citation categories in our analysis scheme.

a large training corpus.

In this study, we report the development of a simplified citation classification schema, a subsequent large annotated corpus, and a deep learning framework for end-to-end ACA.

2 Methods

2.1 Citation Scheme

We developed a simple citation scheme as shown in Table 1. Following Jochim and Schütze (2012), we defined both function classification and sentiment classification schemes as separate facets. For function classification, we followed the widely adopted rhetorical IMARD categories in the scientific domain (Day and Gastel, 2012; Sollaci and Pereira, 2004), and introduced *background*, *method* and *Results/findings* types. We defined the standard *negational*, *confirmative* and *neutral* categories for sentiment classification. We added a *don't know* category to both function classification and sentiment classification since a previous work shows that such a category improved annotation quality (van Rooyen et al., 2015).

2.2 Machine Learning Approaches

We develop deep neural models and compare them with a baseline model for automated citation analysis.

2.2.1 Long Short Term Memory

Long short-term memories (LSTMs) based models are variations of recurrent neural nets and have been introduced to solve the gradient vanishing problem (Hochreiter, 1998). It has an ability to model long-term dependencies of a word sequence (or context) and has achieved notable success in a varity of NLP tasks like machine translation (Sutskever et al., 2014), speech recognition (Graves et al., 2013) and textual entailment recognition (Bowman et al., 2015). In the context of citation analysis, LSTMs read citation context to construct a dense vector representation of the citation for classification.

Let x_t and h_t be the input and output at time step t. Given sequence of input tokens $x_1, \ldots, x_l$ (l is the number of tokens in input text) an LSTM with hidden size k computes a sequence of the output states $h_1, \ldots, h_l$ as

$$i_t = \sigma(W_1 x_t + W_2 h_{t-1} + b_1) \tag{1}$$

$$i'_t = \tanh(W_3 x_t + W_4 h_{t-1} + b_2) \tag{2}$$

$$f_t = \sigma(W_5 x_t + W_6 h_{t-1} + b_3) \tag{3}$$

$$o_t = \sigma(W_7 x_t + W_8 h_{t-1} + b_4) \tag{4}$$

$$c_t = f_t \odot c_{t-1} + i_t \cdot i'_t \tag{5}$$

$$h_t = o_t \odot \tanh(c_t) \tag{6}$$

where $W_1, \ldots, W_8 \in R^{k \times k}$ and $b_1, \ldots, b_4 \in R^k$ are the training parameters. σ and $\odot$ denote the element-wise sigmoid function and the element-wise vector multiplication. The memory cell c_t and hidden state h_t are updated by reading a word token x_t at a time. The memory cell c_t then learns to remember the contextual information that are relevant to the task. This information is then provided to the hidden state h_t by using a gating mechanism and the last hidden state h_l summarizes the all relevant information. i_t, f_t and o_t are called gates. Their values are defined by non-linear combination of the previous hidden state h_{t-1} and the current input token x_t and range from zero to one. The input gate i_t controls how much information needs to flow into the memory cell while the forget get f_t decides what information needs to be erased in the memory cell. The output o_t finally produces the hidden state for the current input token. The final representation vector h_l is subsequently given to a multi-layer perceptron (MLP) with $softmax$ output layer for classification.

Bi-directional LSTMs read the input sequence in both forward and backward directions and have shown to improve further NLP tasks (Jagannatha and Yu, 2016). We implemented Bi-directional LSTM models for citation classification; here we concatenate the last vector representations of the two LSTMs for the subsequent layers.

Studies have shown that LSTMs based models do not work well on memorizing long sequences (Bahdanau et al., 2015). To overcome this limitation, we introduce the attention models.

2.2.2 Global Attention

Attention mechanisms allow NN models to selectively focus on the most task-relevant part of input sequence. As a result, rather than treating every input vector equally, attention models assign weights to the vectors. Since attention models are able to bring out a past and possibly distant input vector to current time step with the blending operation, it also mitigates the information flow bottleneck in RNNs.

We extend the LSTMs based models with a global attention mechanism. This type of attention mechanism is implemented by a neural network that takes a sequence of vectors (usually output vectors of LSTMs) and selectively blends those vectors into a single attention vector. We adopt the attention architecture proposed by Hermann et al. (2015).

Concretely, the global attention considers all the output vectors $h_1, \ldots, h_l$ to construct an attention weighted representation of the input sequence. Let $S \in R^{k \times l}$ be a matrix of the LSTMs output vectors $h_1, \ldots, h_l$ and $o_l \in R^l$ be a vector of ones. An attention weight vector α, an attention representation

r and the final representation h' are defined as

$$M = \tanh(W^a S + W^h h_l \otimes o_l) \qquad (7)$$

$$\alpha = softmax(w^\top M) \qquad (8)$$

$$r = S\alpha^\top \qquad (9)$$

$$h'_l = \tanh(W^s r + W^x h_l) \qquad (10)$$

where $W^a, W^h, W^s, W^x \in R^{k \times k}$ are learnable matrices and $w^\top$ is transpose of learnable vector $w \in R^k$. With the outer product $W^h h_l \otimes o_l$ we repeat the transformed vector of h_l l times and then combine the resulting matrix with the projected output vectors.

2.2.3 Compositional Attention Network

The global attention introduced in the previous section does not incorporate subsequence information as it considers the whole input as a single component. However, natural language and its text form are composed of a set of semantic units. For example, a document can be broken down into paragraphs, the paragraphs into sentences, and the sentences into words. Inspired by this, we propose our CAN model. The proposed attention is also hierarchical in a sense that it consists of different attention layers. CAN attends locally over the input subsequences and globally over the whole input and selectively composes these two types of attention representations with a second layer attention to construct a higher level representation. We use the standard neural attention network (Equation (8 - 10)) from the previous section as a main building block in our CAN.

Let $R \in R^{k \times z}$ be a matrix of the representations $r_1, \ldots, r_z$ (z is the number of input subsequences, i.e. the number of sentences in the input) learned by local attentions and r the output of the global attention. We then obtain the final attention representation r' and the final output h'' as follows.

$$M' = \tanh(W'^a S + W'^h r \otimes o_z) \qquad (11)$$

$$\alpha' = softmax(w'^\top M') \qquad (12)$$

$$r' = R\alpha'^\top \qquad (13)$$

$$h'' = \tanh(W'^s r' + W'^x r) \qquad (14)$$

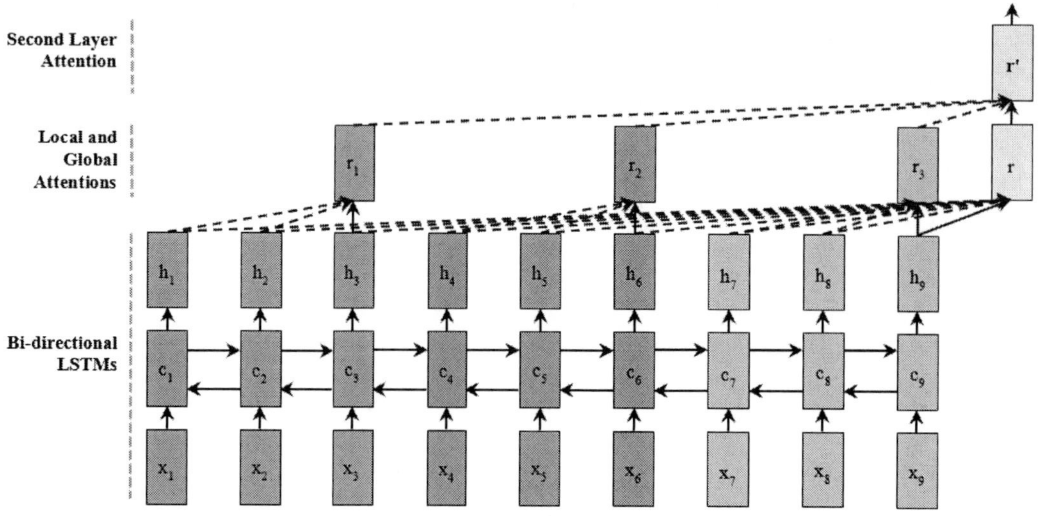

Figure 1: Compositional attention network. r_1, r_2 and r_3 are the locally attended vectors of the output subsets and r is the globally attended vector of the whole output. In the second layer attention, we selectively blend these vectors and obtain the higher level representation r'.

The W matrices and the w vectors of this model can be tied together. When tied, the number of parameters is equal to that of the global attention models. Therefore, this attention network introduces no parametric complexity to compare with the classic global attention model. Figure 1 depicts the overall structure of this model (Equation (1–6), (8–9) and (12–14)). The input consists of the three subsequences $[x_1, x_2, x_3]$, $[x_4, x_5, x_6]$ and $[x_7, x_8, x_9]$. The local attention vectors r_1, r_2 and r_3 are constructed by attending over the LSTM outputs for the each subsequence. Similarly, the global attention vector r is obtained by attending over the whole output sequence $h_1, \ldots, h_9$. In the second layer attention, these representation are composed for the higher level representation r'. The final representation h'' can be obtained according to Equation (9).

The intuition behind our CAN is to attentively compose words within a sentence to construct a local attention vector for each sentence and then these attention vectors are further composed in a second layer attention to learn a whole document representation. We tie the parameters of local, global and the second layer attentions so CAN is forced to learn to compose both the word and sentence presentations attentively.

We also build the bi-directional variation of these models by feeding the concatenated outputs of the forward and backward LSTMs. Due to the concatenated outputs, the size of the W matrices and w vector become $2k \times 2k$ and $2k$ respectively, increasing the number of parameters to be trained.

2.2.4 Baseline Classifier

We implemented a baseline model, which includes extraction of *TF-IDF* statistics of *n-grams* (*1, 2* and *3-grams*) from each citation for feature sets and a support vector machine (SVM) classifier with a linear kernel. For the SVM model, we performed a grid search over its hyper-parameters (including the regularization parameter, C) by using the development set for evaluation. Once the best parameters were found, the final SVM model was learned on both the training and development sets and tested on the test set.

2.3 Data Collection, AMT Annotation and Gold Standard Datasets

In order to increase the generalization of data, we maximizes the total number of selected articles. Specifically, we selected a total of 5,000 citation sentences from 2,500 randomly selected PubMed Central articles (we randomly selected two citation

Corpus	#docs	Avg. #sents	Max. #sents	#classes	Class Distribution
Yelp 2013	335,018	8.9	151	5	.09/.09/.14/.33/.36
IMDB	348,415	14.02	143	10	.07/.04/.05/.05/.08/.11/.15/.17/.12/.18

Table 2: Statistics for the document-level sentiment datasets.

sentences from each article). We then developed guidelines and deployed an annotation task in a crowdsourcing platform, Amazon Mechanical Turk (AMT).

Each citation was labeled by five annotators. We provide the AMT annotators the previous and the next sentences of the citation sentence to enrich the context. We designed a quality control (attention check questions) and ended the AMT session if the AMT workers failed to answer correctly the attention check questions. To evaluate the quality of annotation, we asked a domain expert (a MD) to independently annotate 100 citation sentences randomly selected from our corpus and used it as the gold standard to evaluate inter-annotator agreement with the AMT workers.

We built two gold standard datasets to use for training and for evaluation. The first dataset is composed of labels agreed by at least three of the five annotators (three label matching). This resulted in 3,422 citations for the function analysis and 3,624 citations for the sentiment analysis. The second dataset is more relaxed in which we selected a label given by the majority of the five annotators. In this setting, we included a label that may fail inclusion by the first approach. For example, even if only two annotators agreed on a label, we will include it in our gold standard dataset because it represents a clear majority vote (the rest of three labels all differ). As a result, this dataset included 4,426 citations for

the function classification and 4,423 citations for the sentiment classification.

2.4 CAN for Document-level Sentiment Analysis

In order to test the robustness of the CAN model, we also evaluate it for sentiment analysis on two publically available large-scale datasets: the IMDB movie review and Yelp restaurant review datasets. Particularly, we used the pre-split datasets by Tang et al. (2015). Each document in the datasets is associated with human ratings and we use these ratings as gold labels for sentiment classification. Table 2 reports the statistics for the datasets.

2.5 Experimental Settings

During the experiment, citations labeled with *don't know* were removed from the training data. Each dataset was split into 200/200/rest for dev/test/train sets with a stratified sampling. A stratified sampling is performed to preserve percentage of the citations for each class in each set. We experimented with using only the citation sentence as input example and the expansion with both the previous and the next sentences.

We used ADAM (Kingma and Ba, 2014) for optimization of the neural models. The size of the LSTM hidden units was set to 200. All neural models were regularized by using 20% input and 30% output dropouts and an l_2 regularizer with strength value 1e-3. A word2vec (Mikolov et al., 2013)

Citation Analysis Task	Class	Citation Distribution	
		Majority Voting	Three Label Matching
Function classification	Background	30.5%	20.5%
	Method	23.9%	18.2%
	Results/findings	45.3%	38.3%
	Don't know	0.1%	0.06%
Sentiment classification	Negational	4.8%	2.6%
	Confirmative	75%	59.8%
	Neutral	19.8%	19%
	Don't know	0.2%	0.1%

Table 3: Statistics for our automated citation analysis corpus.

Model	Majority Voting		Three Label Matching	
	Train	Test	Train	Test
SVM	**99.19**	54.27	**87.5**	53.89
LSTMs	59.71	59.55	63.05	66.42
LSTMs + Global Attention	69.02	**65.73**	69.05	**68.61**
Bi-LSTMs	62.14	64.04	67.31	67.88
Bi-LSTMs + Global Attention	72.58	64.6	67.4	**68.61**

Table 4: Citation function classification results. Single **citation sentence** is presented as input.

model trained on a collection of PubMed Central documents transformed citation context to word vectors with size of 200 (Munkhdalai et al., 2015). The parameters of CAN are tied and equal to that of the global attention. The neural models were trained only on the training set while SVM model was built on both training and development sets. We use the development set to evaluate the neural models for each epoch to choose the best model. Each model was given 30 epochs, which was empirically found to be enough time for the models to converge to an optima. The final performances of the methods were reported on the test set. The average training time for the neural network models was approximately three hours on a single GPU (GeForce GTX 980).

3 Results

Table 3 lists the detailed statistics of our AMT annotated corpus. The overall agreement between the expert's annotation and the AMT annotation was 63.1% and 64.7% for function and sentiment analysis tasks. For the function classification, a majority of citations were annotated as results and findings. As shown in Table 3, for the sentiment classification, 4.8% was labeled as *Negational* while 75% and 19.8% were *Confirmative* and *Neutral*. This shows that the citations bias towards a positive statement, resulting a highly unbalanced class distribution.

3.1 Citation Function Analysis

Table 4 lists the results of the function classification by using only citation sentences as input to the models. The SVM baseline obtains the lowest training error. As the models become complex the performance increases. However, some cases like the Bi-LSTMs based global attention model tend to overfit the training data. The unidirectional LSTMs with global attention achieves the best F1-score in both settings when only the citation sentence is input.

Table 5 shows the performance where the inputs are represented by a larger context of the previous, citation and next sentences. We treated the each sentence related to a citation as a subsequence and applied our CAN. Here the bi-directional LSTMs with CAN is the clear winner in terms of the test performance. This model achieves 75.86% F1-score improving the results of the previous model by nearly 7% in the three label matching setup. Unlike the compositional models, the performance of the global attention models decreased in response to additional context given in the input. Furthermore, the models tend to get a higher F1-score in the three label matching setup because this setting has an extra annotation noise filter in selecting the gold labels.

Model	Majority Voting		Three Label Matching	
	Train	Test	Train	Test
SVM	**81.19**	45.72	**99.97**	58.44
LSTMs	53.08	56.74	61.4	59.85
LSTMs + Global Attention	58.97	57.48	77.38	64.96
LSTMs + CAN	60.55	60.11	66.64	73.28
Bi-LSTMs	55.56	56.17	74.49	67.88
Bi-LSTMs + Global Attention	60.28	56.88	66.7	66.42
Bi-LSTMs + CAN	71.34	**60.67**	79.76	**75.57**

Table 5: Citation function classification results. **Citation sentence + its left and right sentences** are used as input.

Model	Majority Voting		Three Label Matching	
	Train	Test	Train	Test
SVM	**82.89**	74.5	**96.39**	70.73
LSTMs	75.25	75.14	73.45	74.48
LSTMs + Global Attention	76.24	76.27	77.38	**75.86**
Bi-LSTMs	75.84	75.7	73.78	75.17
Bi-LSTMs + Global Attention	75.65	**77.4**	80.77	74.48

Table 6: Citation sentiment classification results. Single **citation sentence** is presented as input.

Model	Majority Voting		Three Label Matching	
	Train	Test	Train	Test
SVM	**82.87**	75	**85.72**	71.95
LSTMs	75.25	75.7	73.54	73.79
LSTMs + Global Attention	76.14	75.14	74.68	74.48
LSTMs + CAN	79.18	**76.04**	73.78	**78.1**
Bi-LSTMs	76.24	75.7	73.57	73.79
Bi-LSTMs + Global Attention	75.52	75.7	74.8	74.48
Bi-LSTMs + CAN	75.5	75.44	74.51	75.18

Table 7: Citation sentiment classification results. **Citation sentence + its left and right sentences** are used as input.

3.2 Citation Sentiment Classification

Table 6 shows the evaluation results when the citation sentences are the input. The LSTMs based global attention models obtain the best F1-scores on the test sets. In Table 7, we report the results of the wider context input (citation sentence + its left and right sentences). Here the CAN models perform the best. Similar to the function classification results, the extra context information provides an increasing performance if the model is able to properly exploit.

Despite the same number of training parameters, our compositional attention mechanism significantly improved the performance.

3.3 CAN for Document-level Sentiment Analysis

Table 8 lists our document-level sentiment analysis result on the restaurant and movie review datasets. The CAN model achieves a state-of-the-art by locally and globally composing sentences with its hierarchical attention. The Conv-GRNN and LSTM-GRNN are the best-performing models from Tang et al. (2015)'s and are stacked models of convolutional network and RNNs. Our attention models achieve lower MSEs than the stacked models.

We also analyzed whether lengths influence the performance. We split the Yelp dataset into train/dev/test so the models see only documents with

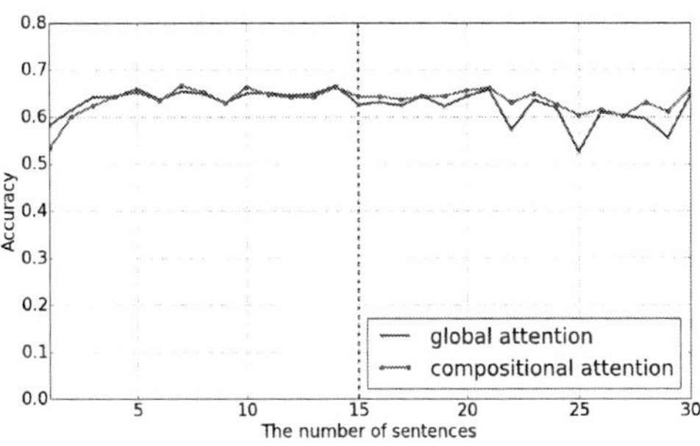

Figure 2: Result on varying length-documents.

length up to 15 sentences during training and classifies much longer documents with length up to 30 sentences during test. Figure 2 plots the test performance over different lengths. The two attention models perform identically on seen lengths except that the global attention model obtains a performance gain on the shorter documents with up to five sentences. However, for unseen lengths (the right side of the green line) the performance of the compositional attention network remains almost consistent and in contrast the global attention starts to decrease in general. This shows the compositional ability of our neural net.

Model	Yelp 2013		IMDB	
	Accuracy	**MSE**	**Accuracy**	**MSE**
SVM (Tang et al., 2015)	59.8	0.68	40.5	3.56
Conv-GRNN (Tang et al., 2015)	63.7	0.56	42.5	**2.71**
LSTM-GRNN (Tang et al., 2015)	**65.1**	**0.5**	**45.3**	3.0
LSTMs + Global Attention (Ours)	63.82	0.57	38.82	**2.25**
LSTMs + CAN (Ours)	**64.49**	**0.55**	**44.16**	2.5

Table 8: Results of document-level sentiment classification. MSE: mean squared error (lower is better).

4 Conclusion

We have developed a generic and simple categorization scheme and a new benchmark corpus for automatic citation analysis. We presented several neural attention networks for the task and evaluated them by using the benchmark corpus. Among these attention mechanisms our original model, we called compositional attention network, performed consistently well on both citation function and citation sentiment classification tasks by attentively composing additional contextual information provided. In an extended experiment, we have also shown that the compositional attention network generalizes better to examples with unseen longer lengths thanks to its compositional operation.

Acknowledgments

We thank the anonymous reviewers for their comments and suggestions. This work was supported in part by the grant HL125089 from the National Institutes of Health (NIH). Any opinions, findings and conclusions or recommendations expressed in this material are those of the authors and do not necessarily reflect those of the sponsor.

References

Shashank Agarwal, Lisha Choubey, and Hong Yu. 2010. Automatically classifying the role of citations in biomedical articles. In *Proceedings of American Medical Informatics Association Fall Symposium (AMIA), Washington, DC*, pages 11–15. Citeseer.

Awais Athar and Simone Teufel. 2012. Context-enhanced citation sentiment detection. In *Proceedings of the 2012 Conference of the North American Chapter of the Association for Computational Linguistics: Human Language Technologies*, pages 597–601. Association for Computational Linguistics.

Awais Athar. 2011. Sentiment analysis of citations using sentence structure-based features. In *Proceedings of the ACL 2011 student session*, pages 81–87. Association for Computational Linguistics.

Dzmitry Bahdanau, Kyunghyun Cho, and Yoshua Bengio. 2015. Neural machine translation by jointly learning to align and translate. In *ICLR*.

Lutz Bornmann and Hans-Dieter Daniel. 2008. What do citation counts measure? a review of studies on citing behavior. *Journal of Documentation*, 64(1):45–80.

Samuel R. Bowman, Gabor Angeli, Christopher Potts, and Christopher D. Manning. 2015. A large annotated corpus for learning natural language inference. In *Proceedings of the 2015 Conference on Empirical Methods in Natural Language Processing, EMNLP 2015, Lisbon, Portugal, September 17-21, 2015*, pages 632–642.

Robert Day and Barbara Gastel. 2012. *How to write and publish a scientific paper*. Cambridge University Press.

Eugene Garfield et al. 1965. Can citation indexing be automated. In *Statistical association methods for mechanized documentation, symposium proceedings*, volume 1, pages 189–92. National Bureau of Standards, Miscellaneous Publication 269, Washington, DC.

Alex Graves, Abdel-rahman Mohamed, and Geoffrey Hinton. 2013. Speech recognition with deep recurrent neural networks. In *Acoustics, Speech and Signal Processing (ICASSP), 2013 IEEE International Conference on*, pages 6645–6649. IEEE.

Marti A Hearst and Emilia Stoica. 2009. Nlp support for faceted navigation in scholarly collections. In *Proceedings of the 2009 workshop on text and citation analysis for scholarly digital libraries*, pages 62–70. Association for Computational Linguistics.

Karl Moritz Hermann, Tomas Kocisky, Edward Grefenstette, Lasse Espeholt, Will Kay, Mustafa Suleyman, and Phil Blunsom. 2015. Teaching machines to read and comprehend. In *Advances in Neural Information Processing Systems*, pages 1684–1692.

Sepp Hochreiter. 1998. The vanishing gradient problem during learning recurrent neural nets and problem solutions. *International Journal of Uncertainty, Fuzziness and Knowledge-Based Systems*, 6(02):107–116.

Abhyuday Jagannatha and Hong Yu. 2016. Bidirectional recurrent neural networks for medical event detection in electronic health records. In *NAACL 2016*.

Charles Jochim and Hinrich Schütze. 2012. Towards a generic and flexible citation classifier based on a faceted classification scheme.

Diederik Kingma and Jimmy Ba. 2014. Adam: A method for stochastic optimization. In *International Conference on Learning Representations*.

Angrosh Annayappan Mandya. 2012. *Enhancing Citation Context based Information Services through Sentence Context Identification*. Ph.D. thesis, University of Otago.

Tomas Mikolov, Ilya Sutskever, Kai Chen, Greg S Corrado, and Jeff Dean. 2013. Distributed representations of words and phrases and their compositionality. In *Advances in neural information processing systems*, pages 3111–3119.

Michael J Moravcsik and Poovanalingam Murugesan. 1975. Some results on the function and quality of citations. *Social studies of science*, 5(1):86–92.

Tsendsuren Munkhdalai, Meijing Li, Khuyagbaatar Batsuren, Hyeon Park, Nak Choi, and Keun Ho Ryu. 2015. Incorporating domain knowledge in chemical and biomedical named entity recognition with word representations. *J. Cheminformatics*, 7(S-1):S9.

Hidetsugu Nanba and Manabu Okumura. 1999. Towards multi-paper summarization using reference information. In *IJCAI*, volume 99, pages 926–931.

Son Bao Pham and Achim Hoffmann. 2003. A new approach for scientific citation classification using cue phrases. In *AI 2003: Advances in Artificial Intelligence*, pages 759–771. Springer.

Luciana B Sollaci and Mauricio G Pereira. 2004. The introduction, methods, results, and discussion (imrad) structure: a fifty-year survey. *Journal of the medical library association*, 92(3):364.

Chris Alen Sula and Matthew Miller. 2014. Citations, contexts, and humanistic discourse: Toward automatic extraction and classification. *Literary and Linguistic Computing*, 29(3):452–464.

Ilya Sutskever, Oriol Vinyals, and Quoc V Le. 2014. Sequence to sequence learning with neural networks. In *Advances in neural information processing systems*, pages 3104–3112.

Duyu Tang, Bing Qin, and Ting Liu. 2015. Document modeling with gated recurrent neural network for sentiment classification. In *Proceedings of the 2015 Conference on Empirical Methods in Natural Language Processing*, pages 1422–1432.

Simone Teufel, Advaith Siddharthan, and Dan Tidhar. 2009. An annotation scheme for citation function. In *Proceedings of the 7th SIGdial Workshop on Discourse and Dialogue*, pages 80–87. Association for Computational Linguistics.

Brendan van Rooyen, Aditya Menon, and Robert C Williamson. 2015. Learning with symmetric label noise: The importance of being unhinged. In *Advances in Neural Information Processing Systems*, pages 10–18.

Hong Yu, Shashank Agarwal, and Nadya Frid. 2009. Investigating and annotating the role of citation in biomedical full-text articles. In *Bioinformatics and Biomedicine Workshop, 2009. BIBMW 2009. IEEE International Conference on*, pages 308–313. IEEE.

Replicability of Research in Biomedical Natural Language Processing:
a pilot evaluation for a coding task

Aurélie Névéol and **Cyril Grouin**
LIMSI, CNRS, Université Paris-Saclay, F-91405 Orsay
`firstname.lastname@limsi.fr`

Kevin Bretonnel Cohen
University of Colorado, USA
`kevin.cohen@gmail.com`

Aude Robert
INSERM-CépiDC, Paris, France
`aude.robert@inserm.fr`

Abstract

The scientific community is facing raising concerns about the reproducibility of research in many fields. To address this issue in Natural Language Processing, the CLEF eHealth 2016 lab offered a replication track together with the Clinical Information Extraction task. Herein, we report detailed results of the replication experiments carried out with the three systems submitted to the track. While all results were ultimately replicated, we found that the systems were poorly rated by analysts on documentation aspects such as "ease of understanding system requirements" (33%) and "provision of information while system is running" (33%). As a result, simple steps could be taken by system authors to increase the ease of replicability of their work, thereby increasing the ease of re-using the systems. Our experiments aim to raise the awareness of the community towards the challenges of replication and community sharing of NLP systems.

1 Introduction

Reproducibility, or replicability, is the quality of a scientific experiment that can be performed independently several times and yield the exact same results on each iteration.

1.1 Why should research strive for reproducibility and methods to achieve it

The advantages of reproducibility notably include increased work productivity and recognition in the community (Piwowar et al., 2007; Schultheiss et al., 2011; Markowetz, 2015). However, in practice, reproducibility is not always achieved or maintained over time (Davis and Walters, 2011). The scientific community is facing raising concerns about the reproducibility of research in many fields (Baker, 2016), including Natural Language Processing (Fokkens et al., 2013).

Is there really a reproducibility problem in natural language processing that needs to be dealt with? Different observations support different conclusions regarding this question. On the one hand, the relative paucity of attention to the question until recently suggests that the community does not seem to think that there is one. On the other hand, recent activity in the area suggests that the community might not be quite so sanguine about the situation: an editorial in a major journal in our field (Pedersen, 2008) and the healthy level of participation in a workshop on the topic associated with a major conference[1] suggest that in fact, reproducibility is an issue—not just reproducibility of work outside of one's own lab, but even reproducibility of work *within* one's own lab.

Can we investigate empirically the extent of reproducibility issues in natural language processing? Previous work has pointed out that in computer science in general, it is difficult to assess reproducibility even at very superficial levels and even with very unambitious definitions of "reproducibility" (Goodman et al., 2016). If the null hypothesis is that it is not any more difficult to assess reproducibility in natural language processing than it is in other areas of computer science, then there is reason to suspect that the null hypothesis does not hold, and that in

---

[1]Workshop on Research Results Reproducibility and Resources Citation in Science and Technology of Language; http://4real.di.fc.ul.pt/

Proceedings of the Seventh International Workshop on Health Text Mining and Information Analysis (LOUHI), pages 78–84,
Austin, TX, November 5, 2016. ©2016 Association for Computational Linguistics

fact it is *more* difficult to assess reproducibility in natural language processing due to the nature of the data our discipline studies: large corpora of natural language texts that are updated on a regular basis (e.g. PubMed, hospital information systems) and subject to being processed in a myriad of different yet similar ways by every researcher (e.g. for the purpose of word segmentation, part of speach tagging).

1.2 The shared task model in evaluation of natural language processing

Early in the history of natural language processing, it was quite difficult for researchers to learn from comparisons of systems because they generally differed on the most basic issues of goals and metrics. Answering questions that are commonplace today, such as *what are the advantages and disadvantages of purely rule-based methods and purely learning-based methods for information extraction?*, was not possible when the differences between projects included not only different methods, but also different extraction targets, data, and figures of merit. In that context, the idea developed that one could learn more from research by standardizing some of those basic aspects of the work. The resulting *shared task model* of evaluation consists of multiple groups agreeing on a shared task definition, a shared data set, and a shared evaluation metric.

Thus, shared tasks provide an opportunity to overcome some of the challenges to replication in natural language processing—in particular, the definitional, data, and scoring issues. The work reported here explores the question of whether the evaluation of replicability in natural language processing can be pushed forward to the highest level of the replicability hierarchy by taking advantage of these aspects of the shared task model. The rationale behind this approach is that one can capitalize on the fact that the systems that are used to address a challenge task all accommodate the same input and output formats, as specified in the challenge. And, the scoring code is open and freely available. Therefore, system results on a challenge dataset should be very easy to replicate without incurring significant training and effort—or, at least, they should be *possible* to replicate, if given access to the original system.

1.3 Leveraging the shared task model to achieve reproducibility

The project reported here is an attempt to pursue the issue of reproducibility in the light of the language-processing-specific problems in research methodology. Discussing reproducibility in computer science in general, Collberg et al. (Collberg et al., 2014) suggest that in the context of computer science research, the notion of reproducibility—defined by them as *the independent confirmation of a scientific hypothesis through reproduction by an independent researcher/lab*—can usefully be replaced by the concept of *repeatability*. In particular, they define three types of what they call *weak repeatability*. The highest level is the ability of a system to be acquired, and then built in 30 minutes or fewer. The next level is the ability of a system to be acquired, and then built, regardless of the time required to do so. The lowest level is the ability of a system to be acquired, and then either built, regardless of the time required to do so, *or* the original author's insistence that the code would build, if only enough of an effort were made.

Previous work has reached only as far as the 3rd level (Anda et al., 2009). However, the shared task environment (defined below) gives us the opportunity to come quite close to the fourth and highest level: the ability of a system to be acquired, built, and used to produce results consistent with published reports. In particular, the facts that the shared task model gives one access to the same data on the one hand, and the same scoring script on the other, provide a rather unique opportunity to evaluate reproducibility at the fourth level. In fact, this paper reports the only work that we are aware of that travels this high up the computer science reproducibility hierarchy.

Reproducibility is a real challenge because of the complexity of scientific experiments and experimental set-up. When describing experiments, researchers are often encouraged to focus on the novelty and interest of their research while devoting less time (and report space) to describe steps that might appear as easy routine. This situation leaves researchers (the authors themselves, or colleagues) trying to reproduce experiments described in a paper with a series of minute technical questions. Without answers to these questions, the experimental set-up may or may not be reproduced exactly, and it becomes difficult to interpret differences in results.

Beyond the observation that reproducibility can be hard to achieve, the scientific community is also trying to understand the specific challenges associated with reproducibility in order to devise strategies to overcome them (Nosek et al., 2015; Cohen et al., 2016). The work we present here follows this direction and aims to study the ease of reproducing experiments in the highly constrained setting of a community shared task, and to yield first-hand actionable knowledge of what makes an experiment easy or difficult to reproduce.

2 Replication track at CLEF eHealth 2016

The CLEF eHealth 2016 lab (Kelly et al., 2016) offered three tasks to promote information extraction and information retrieval in the clinical domain. Task 2 (Neveol et al., 2016) focused on clinical information extraction in languages other than English. It challenged participants with the task of extracting UMLS (Unified Medical Language System) concepts from biomedical text French in the form of anchored normalized entities or ICD10 (International Classification of Diseases, 10th revision) codes.

2.1 Description of the replication task and system requirements

Participation in the replication track was open to all teams who submitted results to the task. After submitting their result files, participating teams had one extra week to submit the system used to produce them, or a remote access to the system, along with instructions on how to install and operate the system.

The "replication track" consisted in attempting to replicate a team's results by running the system supplied on the test data sets, using the team's instructions.

2.2 System analysts and evaluation environment

Four system analysts committed to spend a maximum of one working day (8 hours) with each system. The analysts attempted to install and configure the systems according to the instructions supplied. Participants were also allowed to supply a contact address to make themselves available to address any additional questions.

Two analysts had a Computer Science background with experience developping research systems in the field of bioNLP, and represented the use-case of a colleague trying to reproduce experiments in their field (research-oriented role). Another analyst had a computer science background, and the fourth analyst had a mixed linguistics/computational linguistics background. Both had experience using bioNLP applications and represented the use-case of a user trying to leverage an existing tool for a task of interest (user-oriented role).

In contrast with (Zheng et al., 2015), we did not foster a controlled environment (e.g. using a virtual machine with standard configuration for all analysts) for installing the systems evaluated because we wanted the analysts to work in an experimental setting that would be similar to the one they would use for reproducing experiments. For the same reason, we did not rely on the use of containers.

2.3 Evaluation of the replication experience

The analysts independently ran the systems on the appropriate CLEF eHealth task 2 test sets. The results obtained were be compared to those submitted by the teams using the same system. During this process, the analysts took notes on the various aspects of working with the systems (ease of installing and using, ease of understanding supplied instructions, success of the replication attempt), using a specific score sheet developed by the analysts, following some of the criteria evaluated by (Zheng et al., 2015). The score sheet comprised 10 questions addressing the experience of analysts at each stage of the experiment: system configuration, system installation, running the system, obtaining results, and overall impressions. Table 1 shows the specific questions and answer scales. The analysts were also encouraged to complete their answer to questions with free text comments.

3 Results

A total of seven teams participated in CLEF eHealth 2016 task 2. Three teams submitted systems to the replication track. One team submitted a system that addressed the subtask of named entity extraction and

Question	Scoring Scale
Part 1. System configuration	
Q.1 Is it easy to understand which are the system prerequisites, to check whether they are already installed?	Yes/No
Q.2 Is it easy to follow the installation instructions to install the prerequisites that may be missing?	5-point scale
Part 2. Installing the System	
Q.3 Is it easy to follow the installation instructions to install the system itself?	5-point scale
Q.4 Did you need to contact the system authors to install any part of the system?	Yes/No
Part 3. Running the System on the CLEF eHealth 2016 datasets	
Q.5 Is it easy to follow the instructions in the user manual to use the system to process the challenge dataset(s)?	5-point scale
Q.6 Are there sufficient information to assess whether the system is running as expected, e.g. progress visualization, running time, information messages	Yes/No
Part 4. Obtaining Results	
Q.7 Are the results produced directly in the challenge format?	Yes/No
Q.8 Did applying the challenge evaluation tool yield the exact same results as the participant submitted run?	4-point scale
Part 5. Overall Impression	
Q.9 Do you have any suggestions on what the authors of the system can do to make it more usable? For example: Additional information on where to find prerequisites; Examples of installation or run commands; Screenshots, videos, or tutorials of the installation process or using the system.	free text
Q10. Would you feel comfortable using the system outside the challenge?	Yes/No

Table 1: Score sheet presented to analysts when working with the systems. The 5-point scale comprised the following options: 5-Effortless or nearly effortless, 4-Somewhat easy but there are challenges, 3-Somewhat difficult, 2-Extremely difficult, nearly impossible, 1-I was not able to perform the task. The 4-point scale used for question Q.8 comprised the following options: 4-Yes, exactly the same results, 3- Results are slightly different (less than .01 difference in F-measure), 2- Results are quite different (more than .01 difference in F-measure), 1- Evaluation tool error.

the subtask of ICD10 coding. However, for named entity extraction, the system submitted relied on pre-processed intermediate results obtained by applying an indexing tool on the test corpus. From the perspective of replication, we considered that we had adequate material for reproducing only the ICD10 coding task with this system. The other two systems submitted also addressed the ICD10 coding subtask.

3.1 Characteristics of systems submitted and experimental set-up

Table 2 presents the characteristics of the systems submitted by participants to the replication track. To our knowledge, none of these systems are made available by the authors outside of the CLEF eHealth replication track.

All systems were research prototypes used with terminal-based command-line.

Four analysts (the authors of this paper) participated in the replication experiments. One analyst

Participant	Operating System	Language
System 1	Windows	java
System 2	Linux	python
System 3	Linux	python

Table 2: Characteristics of the systems submitted to the replication track.

had access to both Windows and Linux OS and worked will all three systems. One analyst had access to a Windows OS and worked with System 1. Two analysts had access to a Mac OS and worked with System 2 and 3.

Table 3 presents the configuration of the machines used by the analysts to reproduce experiments.

3.2 Assessment of the replication process

Table 4 presents the time spent by each analyst working with the three systems to attempt reproducing results.

Table 5 presents the aggregated scoring of sys-

Analyst	Configuration
1	Windows 7 16Go ram, 9470Mb cache Intel Core i5-3437U CPU 1.90 GHz (2.40GHz)
2	Windows and Ubuntu 4Go ram, 3MB cache Intel Core i5-3210M with dual-core (2.50GHz)
3	Ubuntu 14.04.3 LTS 62Go ram, 42G cache Intel Xeon, CPU L5520 (2.27GHz)
4	Mac OS X 8Go ram, 3Mb cache Core i5-3427U CPU (1.8GHz)

Table 3: Configuration of the machines used by the analysts to reproduce experiments.

Participant	Analyst	Human Time	Run Time
System 1	User	47	150
System 1	Developer	180	510
System 2	User	204	720
System 2	Developer	45	96
System 3	User	55	240
System 3	Developer	10	93

Table 4: Time (in minutes) spent by each analyst reproducing results with the participant systems. For analysts with the *User profile, human time is averaged between the two analysts, while run time only reflect the run time of the analyst who succeeded in obtaining results from the systems.*

tems performed by analysts while reproducing results.

Phase	Question	Score
Configuration	Q1(*) Easy to understand?	33%
	Q2 Easy to configure?	55%
Installation	Q3(+) Easy to install?	93%
	Q4(*) Contact Author?	0%
Running	Q5(+) Easy to run?	55%
	Q6(*) Info while running?	33%
Results	Q7(*) Challenge format?	100%
	Q8(*) Reproduced?	71%
Overall	Q10(*) Use outside challenge?	33%

Table 5: Aggregated scoring of systems. A star symbol * indicates binary scales (yes/no) and a plus symbol + indicates a 4 or 5 level scale as detailed in table 1. For questions Q7 and Q8, data is averaged over analysts who did succeed in obtaining results.

3.3 Reproducibility of the results

Between them, the analysts were able to replicate results exactly for System 1 and System 3: the precision, recall and F-measure obtained from running the systems were identical to that of the runs submitted by participants for two analysts, while one analyst did not suceed in obtaining results. For System 2, only one analyst was able obtain results (one analyst obtained a memory error before obtaining results and one analyst was not able to run the system), and the results obtained showed a 0.02 difference in F-measure, which was statistically significant. For System 3, one analyst obtained results that showed less than 0.01 difference in F-measure with the results submitted by the participants. For System 3, it can also be noted that the system came with two configuration options and the analysts were only able to implement one of the configuration options each (not the same one). These difficulties are reflected in the score of 71% for the overall reproducibility (Table 5, Q8).

4 Discussion

Almost half of the participants to the CLEF eHealth 2016 task 2 submitted a system to the replication track, and an additional two teams expressed interest in submitting but did not do so due to lack of time and resources to prepare a system suitable for sharing. It can also be noted that the three systems submitted addressed the task of ICD10 coding viewed as a classification task - a relatively simpler task compared to named entity recognition and normalization also offered in task 2, which did not receive any system submission. This confirms that there is a strong interest from the community in the production of reproducible research. It also confirms that the time and resources required to ensure reproducibility are not readily available.

Table 5 shows that the weakest aspects of the systems were the ease of understanding the system requirements (Q1, overall score of 33%) and the quality of information supplied when the systems are actually running (Q6, overall score of 33%). The analysts experienced varying degrees of difficulty to install and run the systems. Differences were mainly due to the technical set-up of the computers used to replicate the experiments. For example, for System

1, one of the analysts had a version of java compatible with the system installed by default, so that running the system was effortless and the question of the java version required never came up. In contrast, the other analyst had an older version of java installed. Running the system produced errors that had to be interpreted to understand that the problem came from the incompatibility between java versions. The analyst then had to look into the system code files to find the java version requirement for the system and then update their work environment accordingly. In our opinion, this highlights the fact that reproducibility needs to be thought through preferably at the time of system development and in any case before a system can be shared or re-used.

Analysts also report that additional information on system requirements, installation procedure and practical use would be useful for all the systems submitted. For system 3, one analyst reported they stopped the experiment because they feared that installing the python configuration required would interfere with the current setting they had and would prevent them from using tools they had set-up. Additional explainations of the system requirement would have helped provide a better understanding of whether the system was compatible with an existing configuration. Free-text comments elicited specific recommandations for each of the systems.

Interestingly, one analyst in the user-oriented role reported that they would feel confident using all of the systems outside the challenge, while the other analysts did not.

5 Concluding remarks

In Section 1, we pointed out that prior to the development of the shared task model, there was no way to explore questions such as *what are the advantages and disadvantages of purely rule-based methods and purely learning-based methods for information extraction?*, due to gross differences in task definition, data, and figures of merit. Despite having developed and matured the shared task model in natural language processing, we still cannot answer questions like that. The shared task model controls three very important variables: task definition, data and evaluation metrics. However, it leaves an *enormous* number of variables unexplored, and those variables can have a large number of values. Suppose that everyone always used the default settings on every out-of-the-box machine learning package: even in that case, one only knows what the default settings are if one knows which version of the package was used, and that is often not recorded in published papers—we looked at 11 of our own machine learning papers, and found that we had given version numbers only 9% of the time.

Nonetheless, the approach that is described in this paper moves the study of replicability in natural language processing forward quite a bit. Replicating the CLEF eHealth challenge results was feasible, and this is the first paper that we know of that has demonstrated that in computer science in general, and in natural language processing in particular. For each of the three systems studied, we were able to replicate the results exactly or closely.

Not only does this work show that the approach is feasible, but it also shows that the approach is able to find problems—a very different kind of value from validating the lack of problems, and in some ways a more valuable one. The ease of replicating results varied. In particular, it generally was based on the analysts' work environment set-up. Moreover, the work reveals something about replicability in natural language processing that is "actionable," something that can be done to improve the situation: most of the difficulties encountered could be alleviated by additional documentation from system authors.

There is some reason to think that the reproducibility situation in natural language processing may be changing, and for the better. The Association for Computational Linguistics is now allowing extra pages in conference papers for documenting the fine details of system configurations. Meetings like the recent workshop at a major conference in the field—and the CLEF eHealth meeting—are exploring the issues and the opportunities for their empirical investigation. In the context of that change, work such as that reported here moves the conversation further along, to higher levels of reproducibility, and it does uncover issues in that respect. The problem of the difficulty of asking the interesting big questions—*what are the advantages and disadvantages of purely rule-based methods and purely learning-based methods for information extraction?*—due to inability to answer the little questions—*which tokenizer did we use, did they use a linear kernel or a radial kernel, do our run times reflect performance before or after we fixed that bug*—may be closer to being resolved.

References

Bente CD Anda, Dag IK Sjoberg, and Audris Mockus. 2009. Variability and reproducibility in software engineering: A study of four companies that developed the same system. *IEEE Trans. Softw. Eng.*, 35(3):407–429.

Monya Baker. 2016. 1,500 scientists lift the lid on reproducibility. *Nature*, 533(7604):452–4.

Kevin B Cohen, Jingbo Xia, Christophe Roeder, and Lawrence Hunter. 2016. Reproducibility in natural language processing: A case study of two R libraries for mining PubMed/MEDLINE. In *LREC 4REAL Workshop:Workshop on Research Results Reproducibility and Resources Citation in Science and Technology of Language*, pages 6–12. European Language Resources Association (ELRA).

Christian Collberg, Todd Proebsting, Gina Moraila, Akash Sankaran, Zuoming Shi, and Alex M. Warren. 2014. Measuring reproducibility in computer systems research. Technical report, Department of Computer Science, University of Arizona.

Philip M Davis and William H Walters. 2011. The impact of free access to the scientific literature: a review of recent research. *J Med Libr Assoc*, 99(3):208–17.

Antske Fokkens, Marieke van Erp, Marten Postma, Ted Pedersen, Piek Vossen, and Nuno Freire. 2013. Offspring from reproduction problems: What replication failure teaches us. In *Proc of ACL*, pages 1691–1701.

Steven N Goodman, Daniele Fanelli, and John PA Ioannidis. 2016. What does research reproducibility mean? *Science Translational Medicine*, 8(341):ps12.

Liadh Kelly, Lorraine Goeuriot, Hanna Suominen, Aurelie Neveol, Joao Palotti, and Guiddo Zuccon. 2016. Overview of the CLEF eHealth evaluation lab 2016. In *Lecture Notes in Computer Science (LNCS), CLEF 2016 7th Conference and Labs of the Evaluation Forum*, Berlin, Heidelberg. Springer.

Florian Markowetz. 2015. Five selfish reasons to work reproducibly. *Genome Biol*, 8(16):274.

Aurelie Neveol, Kevin B Cohen, Cyril Grouin, Thierry Hamon, Thomas Lavergne, Liadh Kelly, Lorraine Goeuriot, Gregoire Rey, Aude Robert, Xavier Tannier, and Pierre Zweigenbaum. 2016. Clinical information extraction at the CLEF eHealth evaluation lab 2016. In *CLEF 2016 Working Notes*, number 609 in CEUR-WS, pages 28–42.

BA Nosek, G Alter, GC Banks, D Borsboom, SD Bowman, SJ Breckler, S Buck, CD Chambers, G Chin, G Christensen, M Contestabile, A Dafoe, E Eich, J Freese, R Glennerster, D Goroff, DP Green, B Hesse, M Humphreys, J Ishiyama, D Karlan, A Kraut, A Lupia, P Mabry, TA Madon, N Malhotra, E Mayo-Wilson, M McNutt, E Miguel, EL Paluck, U Simonsohn, C Soderberg, BA Spellman, J Turitto, G VandenBos, S Vazire, EJ Wagenmakers, R Wilson, and T Yarkoni. 2015. Scientific standards. promoting an open research culture. *Science*, 348(6242):1422–5.

Ted Pedersen. 2008. Empiricism is not a matter of faith. *Computational Linguistics*, 34(3):465–470.

Heather A Piwowar, Roger S Day, and Douglas B Fridsma. 2007. Sharing detailed research data is associated with increased citation rate. *PLoS One*, 2(3):e308.

Sebastian J Schultheiss, Marc-Christian Münch, Gergana D Andreeva, and Gunnar Rätsch. 2011. Persistence and availability of web services in computational biology. *PLoS One*, 6(9):e24914.

Kai Zheng, VG Vinod Vydiswaran, Yang Liu, Yue Wang, Amber Stubbs, Özlem Uzuner, Anupama E Gururaj, Samuel Bayer, John Aberdeen, Anna Rumshisky, Serguei Pakhomov, Hongfang Liu, and Hua Xu. 2015. Ease of adoption of clinical natural language processing software: An evaluation of five systems. *J Biomed Inform*, 58:Suppl:S189–96.

NLP and Online Health Reports:
What do we say and what do we mean?

Nigel Collier
University of Cambridge, Cambridge, UK
nhc30@cam.ac.uk

Abstract

Social media sites such as microblogs and discussion board forums have the potential to be rich source of information about human health. Going beyond simple keyword search and harnessing the data for insights that can benefit public health presents both opportunities and challenges to natural language processing (NLP). In this talk I survey the progress made using NLP methods, e.g. for adverse drug reaction profiling, flu surveillance and the study of depressive disorders. I will then look at the technical challenges in understanding such messages, in particular how NLP can automatically encode/normalise laymen's language to the formal terminologies of healthcare professionals. To this end I present state of the art results from our recent work on using deep neural networks to de-conflate word senses as well as 'translating' from social media messages to SNOMED-CT. I will finish by briefly reflecting on the practical and ethical challenges that lie ahead.

Speaker bio

Nigel is PI and Director of Research in Computational Linguistics at the Department of Theoretical and Applied Linguistics in the University of Cambridge. Nigel was awarded a PhD in computational linguistics from the University of Manchester (UMIST) in 1996 for his work on Lexical Transfer using a Hopfield Neural Network. He was awarded a Toshiba Fellowship to continue his research on neural networks for machine translation and then joined the NLP group at the University of Tokyo where he coordinated the GENIA text mining project. After becoming faculty at the National Institute of Informatics (NII) in 2000, Nigel led the BioCaster research programme (2006 to 2012) for multilingual news surveillance and served as technical advisor to the Global Health Security Action Group's working group on Risk Management and Communication. He was awarded a Marie Curie fellowship at the European Bioinformatics Institute from 2012 to 2014 where he continued his investigation into biomedical text mining for scientific texts.

Nigel's research interests are in information extraction and biomedical knowledge discovery with a focus on machine learning approaches for representation learning of concepts. He is the author of over 90 peer-reviewed articles and conference papers on biomedical NLP. Nigel currently leads the EPSRC-funded Semantic Interpretation of Personal Health messageS (SIPHS) project which investigates biomedical concept encoding of laymens terms in the social media for real world applications such as digital disease surveillance (https://sites.google.com/site/nhcollier/).

Proceedings of the Seventh International Workshop on Health Text Mining and Information Analysis (LOUHI), page 85,
Austin, TX, November 5, 2016. ©2016 Association for Computational Linguistics

Leveraging coreference to identify arms in medical abstracts: An experimental study

Elisa Ferracane
Department of Linguistics
The University of Texas at Austin
elisa@ferracane.com

Iain Marshall
Dept. of Primary Care and Public Health Sciences
Kings College London
iain.marshall@kcl.ac.uk

Byron C. Wallace
College of Computer and Information Science
Northeastern University
byron@ccs.neu.edu

Katrin Erk
Department of Linguistics
The University of Texas at Austin
katrin.erk@mail.utexas.edu

Abstract

Performing systematic reviews is a critical yet manual, labor-intensive step in evidence-based medicine. Automating systematic reviews is an active area of research, requiring innovations in machine learning and computational linguistics. We examine how coreference resolution can aid in identifying the arms of a study, an often overlooked piece of information needed to synthesize the results in a systematic review. A classification model[1] that performs better with the coreference features supports the intuition that coreference is able to capture the discourse salience of arms. We note that control arms do not benefit as much from these features.

1 Introduction

Evidence-based medicine (EBM) is a paradigm that seeks to inform medical practitioners of the optimal treatment, based on the totality of the available evidence (i.e., the results of all relevant clinical trials). To this end, teams of medical experts often conduct *systematic reviews*, which synthesize all published medical literature pertaining to a specific clinical question. The first step in a systematic review is to formulate the research question to be investigated, and then find all of the relevant citations. Abstracts and then full texts are screened to exclude irrelevant trials. Once a set of trials pertinent to the research question are identified (typically 10-20 trials), key pieces of information are extracted from each trial. This information generally consists of the patient Population under study, the Intervention(s) being tested, the Comparison and the Outcomes (abbreviated as PICO). Results from all identified trials are typically statistically combined via meta-analysis to produce an aggregated result.

Producing systematic reviews is a time-consuming, largely manual process. This is exacerbated by the rapidly growing evidence base: PubMed[2] contains 800,000+ publications on clinical trials in humans (Wallace et al., 2013), and on average reports of 75 new trials are published daily. A single systematic review can take over a year to produce – at which point it risks becoming outdated. Therefore, automating evidence synthesis poses an enormous yet enticing challenge for automation.

A crucial step towards automating synthesis is identifying the *arms*, or groups, in trials. A clinical trial consists of one control arm, and one or more intervention arms. For example, a study comparing the efficacy of aspirin versus a placebo would consist of two arms: those taking *aspirin* (the intervention group), and those taking the *placebo* (the control group). Previous work has mostly focused on identifying the PICO elements. However, the PICO elements alone are insufficient to convey the design of the study, a key piece of evidence necessary in the downstream task of data synthesis and analysis. Thus, the present study focuses on improving the automated identification of arms. We observed that arms are often salient in the discourse of the abstract, in that they corefer more often than other to-

---

[1] https://github.com/elisaF/extractGroups

[2] publicly available resource for accessing medical references and abstracts https://www.ncbi.nlm.nih.gov/pubmed/

Proceedings of the Seventh International Workshop on Health Text Mining and Information Analysis (LOUHI), pages 86–95,
Austin, TX, November 5, 2016. ©2016 Association for Computational Linguistics

Randomised controlled trial with 12 month intervention. Change in body mass index (BMI) standard deviation score (SDS) over 12 months with assessment 18 months after the start of the intervention. Using the last available data on all participants (n=106), those in the Mandometer group [arm1] had significantly lower mean BMI SDS at 12 months compared with standard care [arm2]. The mean meal size in the Mandometer group [arm1] fell by 45 g. Those in the Mandometer group [arm1] also had greater improvement in concentration of high density lipoprotein cholesterol.

Table 1: Excerpt from medical abstract illustrating the discourse salience of the intervention arm, *arm1*, where the control arm is *arm2* (note that not all mentions of the arms are annotated in the gold data, as discussed in section 5.3).

Randomised controlled trial with [chain1] 12 month intervention . Change in body mass index (BMI) standard deviation score (SDS) over 12 months with assessment 18 months after the start of [chain1] the intervention . Using the last available data on all participants (n=106), those in [chain2] the [chain3] Mandometer group [arm1] had significantly lower mean BMI SDS at 12 months compared with standard care [arm2]. The mean meal size in the [chain3] Mandometer group [arm1] fell by 45 g. Those in the [chain2] [chain3] Mandometer group also had greater improvement in concentration of high density lipoprotein cholesterol.

Table 2: Medical abstract with annotated arms and coreference chains. The chains were automatically determined as described in section 4.3. All phrases with the same chain label are judged to co-refer.

kens. This study is exploratory work that focuses on investigating the effectiveness of using coreference features for identifying arms.

The remainder of this paper is organized as follows. We motivate the choice of coreference features for arm identification. We then examine prior work in identifying the arms in medical texts, and how coreference resolution has been applied to the medical field. Next, we present an experiment to classify whether tokens in annotated medical abstracts are part of an arm. We propose features that take advantage of the discourse salience of arms, and we discuss the results with and without the coreference features.

2 Motivation

Identifying the arms is not a simple information extraction task. The arms in a study consist of one control group, and one or more intervention groups. Often, the control group is never explicitly mentioned in the abstract. In the following excerpt, only the intervention arm is mentioned:

To determine whether modifying eating behaviour with use of a feedback device facilitates weight loss in obese adolescents.

An arm in a study is typically a noun phrase (NP), where this NP is repeated, either verbatim or anaphorically, throughout the abstract. An example of the discourse salience of arms in a medical abstract is in Table 1. The intervention arm, *Mandometer group*, is repeated several times verbatim throughout the abstract.

Given this recurring linguistic pattern in medical abstracts, we investigated the use of *coreference resolution* to help identify arms. The goal of coreference resolution is to determine which mentions in a text refer to the same entity. A referring expression, or *mention*, is the natural language expression used by discourse participants to refer to entities. Two or more mentions that refer to the same entity are coreferent, and together form a *coreference chain*. An anaphor and its antecedent (or cataphor and its postcedent) will form a coreference chain. Mentions can be indefinite noun phrases, definite noun phrases, proper names and pronouns, where clinical trial abstracts contain mostly NP's. Using an off-the-shelf coreference tool (to be discussed in more detail in section 4.3) yields the mentions and coreference chains illustrated in Table 2.

Note that the token *intervention*, which is not part of an arm, appears at most 2 times within a single coreference chain, whereas *Mandometer*, part of the experimental arm, appears 3 times. Further, *intervention* is found only in 1 chain, whereas *Mandometer* appears in 2 chains. More generally, we hypothesize a token forming part of an arm is more salient in two ways: (i) an arm token appears more often within a single coreference chain, and (ii) an arm token appears more frequently across different chains (within the same abstract). These observations motivate the coreference features presented in section 4.3. In Table 2, *standard care* is not a member of any chains. More generally, we can expect salience to help more with intervention arms than control.[3]

3   Related work

3.1   Automated Identification of Arms

Previous work has identified PICO elements either at the word or sentence level. Most research has extracted information from medical abstracts, although some studies have used the full text of the articles (De Bruijn et al., 2008; Zhao et al., 2012; Wallace et al., 2016). One of the seminal studies in PICO extraction (Demner-Fushman and Lin, 2007) collapsed intervention and comparator, where interventions were short noun phrases based largely on recognition of semantic types (mapped to UMLS concepts) and a few manually constructed rules. The intervention/comparator extractor returned a list of all the interventions under study, and the extractor was evaluated at the sentence level. However, it is important to distinguish between experimental and control treatments as the bias for the experimental

group must be accounted for in the data synthesis step (Lumley, 2002).

Beyond PICO, De Bruijn et al. (2008) extracted data from full-text articles based on the CONSORT Plus Guideline,[4] a list of required, recommended and optional items to include in a systematic review compiled by medical experts. The study found that one of the most difficult items to identify was the experimental treatment, which varied widely beyond just drug names. Elsewhere, Chung (2009) identified interventions as a coordinating structure in a single sentence, and found the major weakness in this approach was parsing errors when identifying the boundaries of the conjuncts. And Summerscales et al. (2011) focused on the downstream task of calculating the absolute risk reduction (ARR), identifying the number of bad outcomes for the control and experimental treatment groups, along with the sizes of both treatment groups. This study found outcomes hardest to detect because of their variability, but also had an overall poor recall partly because coreference was not taken into account.

Most recently, Trenta et al. (2015) proposed a novel approach for identifying the arms and PICO elements that does not rely on a first stage of sentence classification, but instead classifies each token directly, followed by an inference process to constrain the labels to more accurate results. As with previous studies, outcome results were the hardest because they are more variable. A significant limitation of this study is that the abstracts were limited to two-arm trials, and in a specific domain.

3.2   Automated Coreference Resolution

Coreference resolution is a long-studied task that remains a challenging problem. Most recent work on coreference resolution builds mainly on one of four models.

- The first and most widely-used approach is the *mention-pair* model (Soon et al., 2001; Ng and Cardie, 2002b). A classifier first identifies all the pairs of mentions which are coreferent. These pairs are then grouped into coreferent chains by clustering techniques such as closest-first (Soon et al., 2001) or best-first (Ng and Cardie, 2002b; Ng and Cardie, 2002a).

---

[3]Cases of joint coreference such as *all participants* referring to both arms in the example abstract are not addressed in this paper, but pose an interesting problem for identifying PICO elements such as population and outcome.

[4]http://rctbank.ucsf.edu/home/cplus

In closest-first, you link to the closest preceding mention, whereas in best-first, you choose the likeliest one. Common features in these models include distance between the two mentions, syntactic features (e.g., POS tags), semantic features (e.g., named entity type), lexical features (e.g., head word of the mention), and string matching.

- The *mention-ranking* model (Denis and Baldridge, 2008), reframes the task as a ranking function rather than a classification function, ranking all the candidate antecedents of a mention to determine which candidate antecedent is the most probable.

- The *entity-centric* model makes use of entity-level information, focusing on features of mention *clusters*, and not just pairs (Raghunathan et al., 2010). The coreference clusters are built up incrementally, using information from partially-completed coreference chains to guide later decisions. Features include whether a mention head word matches any of the head words in the antecedent cluster.

- The *antecedent tree model* (Yu and Joachims, 2009) builds a graph from a document, where the nodes are the mentions and arcs are the links between mention pairs that are coreferent candidates. The coreference chains are then modeled as latent trees in the graph.

Constraints are imposed on these models for improved results, such as enforcing a transitive closure to guarantee you end up with legal assignments (Finkel and Manning, 2008). For example, if *John Smith* is coreferent with *Smith*, and *Smith* with *Jane Smith*, then it should not follow that *John Smith* and *Jane Smith* are coreferent. Other work has shown that joint models improve performance. Denis et al. (2007) recognized that anaphoricity (whether an entity is the first mention) and coreference should be treated as a joint task since one informs the other. Durrett and Klein (2014) models coreference together with named entity recognition and linking named entities to Wikipedia entities. Combinations of these models have also yielded improved results, such as Clark and Manning (2015) stacking

mention-pair and *entity-centric* systems (which the current paper uses as its off-the-shelf coreference resolver).

Many coreference resolvers exploit deeper linguistic knowledge, beyond the features mentioned above. Chowdhury and Zweigenbaum (2013) eliminated less-informative training instances prior to model training by creating a list of criteria based on semantic and syntactic intuitions such as a mismatch in semantic types. Peng et al. (2015) created predicate schemas to constrain inference, such as two predicates with a semantically shared argument. Yang et al. (2015) used semantic role labeling to link the time and locations for event mentions, and for verbal mentions they linked their participants. More recently, Kilicoglu et al. (2016) focused on sortal anaphoras which they found to commonly occur in biomedical literature, resolving anaphors that carry a specific semantic type, or sort, such as *these drugs*. Many of these studies take advantage of linguistic resources such as WordNet[5] and FrameNet[6].

In the medical area, coreference resolution has been most closely studied for analyzing clinical narrative text such as that found in Electronic Health Records (EHRs), and biomolecular studies. In fact, there have been corpora (i2b2/VA Corpus(Uzuner et al., 2012), GENIA Event Corpus(Kim et al., 2008)) and shared tasks (SemEval-2015 shared task on Analysis of Clinical Text (Task 14)(Elhadad et al., 2015), BioNLP09 shared task(Kim et al., 2009), ShARe/CLEF eHealth 2013 Evaluation Lab Task 1(Pradhan et al., 2013)) created specifically to advance this area. Given that resources such as FrameNet and WordNet are based mostly on news (e.g. British National Corpus, U.S. newswire), a large number of resources have been created to aid in natural language processing of medical texts. By far the largest and most complex is the Unified Medical Language System (UMLS)[7], consisting of three main components: Metathesaurus with terms and codes from many vocabularies (including CPT, ICD-10-CM, MeSH, RxNorm, and SNOMED CT), Semantic Network with semantic types and semantic relations, and the SPECIALIST Lexicon, which contains syntactic, morpholog-

---

[5] http://wordnet.princeton.edu
[6] https://framenet.icsi.berkeley.edu
[7] https://www.nlm.nih.gov/research/umls/

ical and orthographic information on terms, along with NLP tools such as POS tagger and word sense disambiguator. Other tools include MetaMap[8], a tool for recognizing UMLS concepts, DrugBank[9], a database of drug names, BANNER[10], a named entity recognizer for biomedical texts, BioText for identifying entities and relations in bioscience texts, and BioFrameNet[11], an extension of FrameNet for molecular biology (and BioWordNet(Poprat et al., 2008) was a failed attempt at extending WordNet also to the biomolecular field). However, when applied to clinical trial texts, these tools prove useful mainly for identifying only medical terms and drug names, and thus more linguistically-motivated resources are still lacking for clinical trial texts.

In the area of clinical narratives, Raghavan et al. (2012) took advantage of the temporal features present in these texts to help determine whether two medical concepts corefer with each other. Their 2014 paper (Raghavan et al., 2014) expanded on this idea to identify medical events spanning across narratives, such as admission notes, medical reports, and discharge notes. Yoshikawa et al. (2011) exploited coreference information for extracting event-argument relations from biomedical texts in the Genia Event Corpus. Jindal and Roth (2013) used very specific domain knowledge to resolve coreference in clinical narratives, such as creating a specific discourse model (i.e. a single patient, several doctors and a few family members) to resolve entities of type "person". Despite the active interest in coreference resolution, there has been much less research investigating its application to clinical trial texts. Most of the literature that does exist is applied to the bio-medical field, focusing more on full-text articles (Gasperin and Briscoe, 2008; Huang et al., 2010; Kilicoglu et al., 2016) than on abstracts (Castano et al., 2002; Yang et al., 2004). To the best of the authors' knowledge, there have been no papers using coreference features to identify arms in clinical trial abstracts.

---

[8]https://metamap.nlm.nih.gov
[9]http://www.drugbank.ca
[10]http://banner.sourceforge.net
[11]http://biotext.berkeley.edu

4 Experiment

The goal of this experiment is to explore empirically whether incorporating coreference features improves the performance of a classifier for arm identification, as compared to a baseline model without coref features (note that we do not aim to necessarily achieve state-of-the-art results on this task). The task of the classifier is to label a token as either part of an arm or not.

4.1 The corpus

The corpus[12] consists of 263 abstracts from the British Medical Journal (BMJ) annotated with the experimental and control groups (and other PICO elements) by Summerscales (2013). The BMJ requires structured input, and the number of sections varies with some abstracts only containing a few sections such as BACKGROUND, METHODS, FINDINGS and INTERPRETATION. These structured abstracts usually consist of short phrases and incomplete sentences.

Number of documents	263
Number of tokens	63,488
Number of [abstract, token] pairs	35,650
Average no. tokens per document	241
Positive labels	5,757 (9%)

Table 3: Corpus statistics

4.2 Experimental setup

Sentences were tokenized, lower-cased and stop words were removed . Each token was paired with its abstract to form an *[abstract, token]* pair to uniquely correlate the token with the medical abstract where it appeared (e.g. *[abstract_3, "intervention"], [abstract_129, "intervention"]*). A binary classifier was implemented to label each token as belonging to an arm or not (scikit-learn implementation of Support Vector Machine, Pedregosa et al. (2011)). Due to the imbalance of classes (9% positive), the class weights in the model were adjusted to be inversely proportional to the class frequencies in the corpus. We performed five-fold cross validation.

---

[12]https://github.com/rlsummerscales/
bibm2011corpus

Model	Precision (var)	Recall (var)	F1 (var)
baseline	12.9 (2.7e-04)	**88.6** (5.6e-04)	22.5 (6.2e-04)
coref	**19.7** (7.5e-04)	82.7 (8.4e-04)	**31.8** (14.4e-04)

Table 4: Results averaged across 5-folds on the two models with their variances in parentheses.

Feature	Mean	Range	Variance
b-o-w	1.78	1-24	2.71
drugbank	0.09	0-1	0.08
tf-idf	6.06	1-141.1	42.67
coref max_counts	0.14	0-15	0.31
coref num_chains	0.10	0-6	0.11

Table 5: Feature statistics

4.3 Features

The following features, summarized in Table 5, were used in the machine learning algorithm.

bag-of-words The number of times the token occurs within its medical abstract (i.e., the count of *[abstract, token]* pairs for the given token and abstract). As evident in Table 5, abstracts can be quite repetitive in their vocabulary, but on average a token appears only a couple of times within the same abstract.

drugbank Whether the token exists in the Drug-Bank database version 4.3[13]. The clinical trials often compare the efficacy of different drugs, such that intervention arms would contain drug names. However, note from Table 5 that most words are not drugs, keeping in mind that interventions also consist of therapies, behavior changes and other non-drug-related treatments.

tf-idf: Term frequency-inverse document frequency for term t in document d for corpus D:

$$tf\text{-}idf_{t,d} = tf_{t,d} * (idf_{t,D} + 1), \qquad (1)$$

where:

$$tf_{t,d} = f_{t,d}$$

$$idf_{t,D} = \log \frac{|D|}{|\{d \in D : t \in d\}|}$$

One is added in the equation (1) so that terms with zero idf (those that occur in all documents of a training set) are not entirely ignored. The goal of this metric is to capture how informative a word is. For example, the token *mandometer* (an arm) from the abstract in Table 2 has a tf-idf measure of 26.29, whereas *intervention* (not an arm) has a value of 3.7. On average, the tokens are slightly more informative than common words such as *intervention*.

coreference:

The Coreference Resolution annotator packaged in Stanford Core NLP 3.0[14] (a model that stacks *mention-pair* and *entity-centric* systems) is used to calculate the maximum number of times the token occurs in a single coreference chain within the same medical abstract (**max_counts**) and the number of chains the token appears in the same medical abstract (**num_chains**). This tool was chosen because it is publicly available and yields state-of-the-art results on the 2012 CoNLL data set. The coreference features aim to capture the discourse salience of arms in medical abstracts. As mentioned before, the (max_counts, num_chains) values for *mandometer* are (3,2), but for *intervention* are (2,1). Note from Table 5 that although a token can occur very frequently in a single chain (*max_counts*) and across many chains (*max_chains*), a token on average is not part of a chain at all. This observed statistic lends weight to the use of coreference features as a measure of salience. Previous work has employed other features such as dependency trees and other predicate argument structures to capture this discourse salience. Summerscales (2013) implemented a form of post-hoc coreference resolution as a way to cluster labeled words into groups, for example into a control group versus an intervention group. However, the present study uses the coreference features at the front end to detect the mentions, and is presently not concerned with differentiating among the different arms.

---

[13] http://www.drugbank.ca/system/downloads/
4.3/drugbank.xml.zip

[14] http://nlp.stanford.edu/software/
stanford-corenlp-full-2015-12-09.zip

5 Evaluation

Table 4 summarizes the evaluation scores. The results of the classifier are evaluated against the spans of text that were annotated as arms, following Summerscales (2013). Because an arm consists of several contiguous words (e.g. *mandometer group*), we want to ensure the classifier is able to correctly label the more informative words in that span (*mandometer* vs. *group*). A labeled group of words is considered a match for an annotated group if they consist of the same set of words, ignoring *had, group(s)*, and *arm*. For example, a labeled span of *mandometer* for the annotated span *mandometer group* is a true positive. On the other hand, a labeled span of only *group* is a false positive. Although the scores are relatively low for both models, we emphasize the goal of this experiment is not to achieve state-of-the-art results but to investigate the viability of salience for arm identification. Further, we are being strict in our evaluation, compared to prior work (e.g., Summerscales (2013)).

5.1 Baseline

The **baseline** model includes the features for how many times a token appears in a single abstract (**b-o-w**), whether the token exists in the Drug-Bank (**drugbank**), and the term-frequency inverse-document-frequency measure for the token (**tf-idf**).

5.2 With Coreference

The **coref** model additionally includes the maximum number of times the token appears in a single coreference chain for a given abstract (**max_counts**), and the number of coreference chains the tokens appears in for a given abstract (**num_chains**).

5.3 Error Analysis

The coref model performed better than the baseline model in almost all the metrics: precision (improved 6.8 points) and F1 (+9.3). Additionally, these improvements are consistent across all the cross-validation runs, as illustrated in Figure 1. Adding the coreference features lowers recall by 5.9 points. To understand the results in more detail, we compare the confusion matrices of the two models. The raw counts in Figure 2 illustrate the class imbalance of the data, giving the impression that a false positive

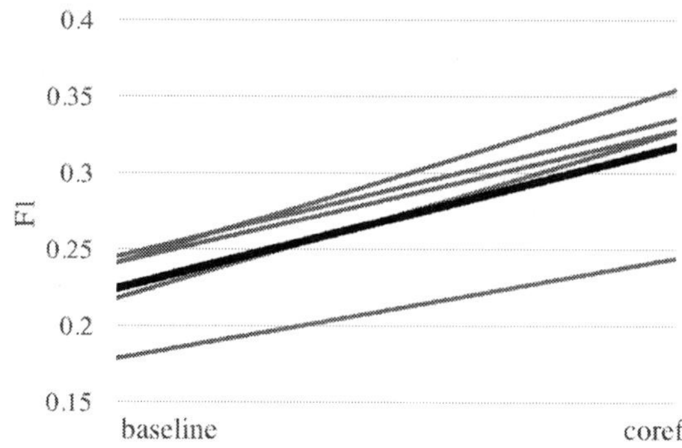

Figure 1: F1 score across the 5 runs in gray, with mean in the thick black line. The lines connect results in the baseline model to results of the the same folds in the coref model.

is more likely than a false negative. The normalized confusion matrices in Figure 3 show that false negatives are a higher percentage of the errors than false positives, so that the positive class is the harder one to label.

Given that false negatives are the most common errors across both models, we analyze their occurrences first. The control arm is the most susceptible to this type of error, as it is not as salient in the discourse as the experimental arms. The control words are typically drawn from a finite and small vocabulary (e.g. *control, placebo, sham, standard*), so their tf-idf scores are usually low. The false negative rate worsens in the coref model partly because it places more weight on discourse salience, and control arms are often not part of a coreference chain, compared with experimental arms. We refer back to the abstract presented in Table 1. A small ablation study was conducted to determine that the b-o-w feature is able to correctly label *standard* (count=4) as part of an arm. With the coreference features, the word is no longer labeled as an arm, as it does not appear in any coreference chain.

Next, we analyze the false positives across both models. Given that all the features (except drugbank) in both models are aimed at extracting salient words, they also pick out other relevant PICO information. For example, both models incorrectly label *knee* as part of an arm in the following abstract, where each of these mentions is, in fact, annotated as part of an *outcome*:

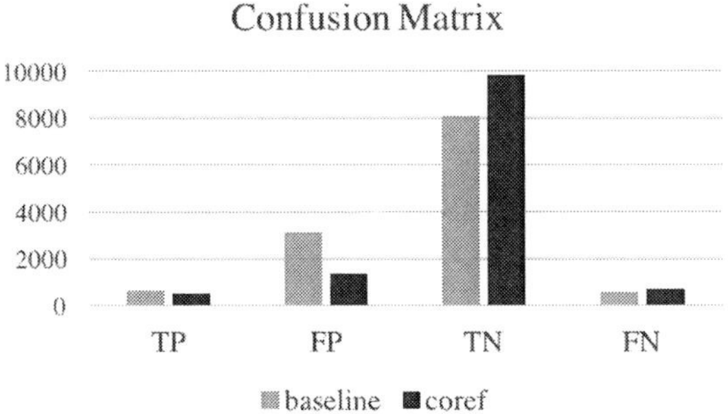

Confusion Matrix

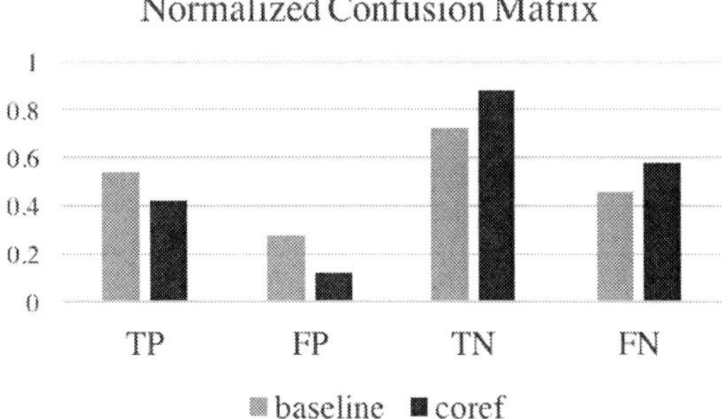

Normalized Confusion Matrix

Figure 2: The raw counts of the confusion matrices for the baseline and coref models.

...reduce the incidence of knee and ankle injuries in young people participating in sports. The rate of acute injuries to the knee or ankle. A structured programme of warm-up exercises can prevent knee and ankle injuries...

Another issue with false positives is that the gold data is not comprehensively annotated. Note that in Table 2, the annotator failed to label the third occurrence of *mandometer* as an arm, although both models attempt to classify it as such. However, striving for a thoroughly annotated data set is not realistic, and so the models should be more robust to these gaps and inconsistencies. The false positive rate improves in the coref model partly because the coreference features prove to be a better measure of discourse salience for the intervention arms. As noted earlier, repetition in medical abstracts is not limited to the words describing the arm. For example, in the abstract from Table 1, the baseline model incorrectly labels the high-frequency tokens *eating, months* and *mean* as parts of an arm. The coref model instead correctly labels these as negative, given that they do not occur in a coreference chain.

Finally, we note that the coreference features help in grouping together words with conflicting tf-idf measures. In the abstract from Table 1, the baseline model correctly labels *mandometer* (tf-idf=26.3), but misses *group* (tf-idf=4.2). However, the coref model correctly labels the entire span *mandometer group* as an arm, because both of these tokens appear together in a mention and have the same coreference features.

6 Conclusion

We introduced a new approach to identify the arms in a clinical trial abstract by creating coreference

Figure 3: The normalized confusion matrices for the baseline and coref models.

features aimed at capturing the discourse salience of arms. The coreference features were shown to help in classifying a word as part of an arm, confirming the intuition that mentions of arms throughout the abstract often corefer. However, we note this pattern holds more for the experimental than control arms. The error analysis also revealed that arms are not the only concepts that are coreferent: other PICO elements such as the outcome often have the same features. This observation could motivate a model that jointly labels these PICO elements along with the arms, since one would inform the other. There are several other recurring linguistic patterns yet to be explored that could further aid in arm identification, such as apposition:

A computerised device, Mandometer, providing real time feedback...

and paraphrasing:

..half were produced automatically with a larger volume of material...The larger booklets produced automatically were...

Another avenue of research is to investigate how these linguistic features pattern across abstracts in the same review. For example, finding the paraphrases across all abstracts that study the same treatment (as defined in a systematic review) could yield finer-grained information on the language used to describe that intervention. To compensate for the inconsistent and small number of annotations, label propagation might be used to retrieve clusters of relations and find the structure in the data.

As noted earlier, the present study focused on the effect of salience on arm identification. In a future study, we plan to implement Summerscales (2013) as a strong baseline (which achieved an F-score of 0.69) to understand whether coreference can still yield improved results when compared to a model that nears state-of-the-art performance.

Acknowledgments

We thank Dr. Rodney Summerscales for providing us with the annotated corpus, and the anonymous reviewers for their helpful feedback.

Wallace and Marshall were supported by the National Library of Medicine (NLM) of the National Institutes of Health (NIH) under award number R01LM012086. The content is solely the responsibility of the authors and does not necessarily represent the official views of the National Institutes of Health.

References

José Castano, Jason Zhang, and James Pustejovsky. 2002. Anaphora resolution in biomedical literature.

Md Faisal Mahbub Chowdhury and Pierre Zweigenbaum. 2013. A controlled greedy supervised approach for co-reference resolution on clinical text. *Journal of biomedical informatics*, 46(3):506–515.

Grace Yuet-Chee Chung. 2009. Towards identifying intervention arms in randomized controlled trials: extracting coordinating constructions. *Journal of biomedical informatics*, 42(5):790–800.

Kevin Clark and Christopher D. Manning. 2015. Entity-centric coreference resolution with model stacking. In *Association for Computational Linguistics (ACL)*.

Berry De Bruijn, Simona Carini, Svetlana Kiritchenko, Joel Martin, and Ida Sim. 2008. Automated information extraction of key trial design elements from clinical trial publications. In *AMIA Annual Symposium Proceedings*, volume 2008, page 141. American Medical Informatics Association.

Dina Demner-Fushman and Jimmy Lin. 2007. Answering clinical questions with knowledge-based and statistical techniques. *Computational Linguistics*, 33(1):63–103.

Pascal Denis and Jason Baldridge. 2008. Specialized models and ranking for coreference resolution. In *Proceedings of the Conference on Empirical Methods in Natural Language Processing*, EMNLP '08, pages 660–669, Stroudsburg, PA, USA. Association for Computational Linguistics.

Pascal Denis, Jason Baldridge, et al. 2007. Joint determination of anaphoricity and coreference resolution using integer programming. In *HLT-NAACL*, pages 236–243. Citeseer.

Greg Durrett and Dan Klein. 2014. A joint model for entity analysis: Coreference, typing, and linking. *Transactions of the Association for Computational Linguistics*, 2:477–490.

Noémie Elhadad, Sameer Pradhan, WW Chapman, Suresh Manandhar, and GK Savova. 2015. Semeval-2015 task 14: Analysis of clinical text. In *Proc of Workshop on Semantic Evaluation. Association for Computational Linguistics*, pages 303–10.

Jenny Rose Finkel and Christopher D. Manning. 2008. Enforcing transitivity in coreference resolution. In *Proceedings of the 46th Annual Meeting of the Association for Computational Linguistics on Human Language Technologies: Short Papers*, HLT-Short '08, pages 45–48, Stroudsburg, PA, USA. Association for Computational Linguistics.

Caroline Gasperin and Ted Briscoe. 2008. Statistical anaphora resolution in biomedical texts. In *Proceedings of the 22nd International Conference on Computational Linguistics-Volume 1*, pages 257–264. Association for Computational Linguistics.

Cuili Huang, Yaqiang Wang, Yongmei Zhang, Yu Jin, and Zhonghua Yu. 2010. Coreference resolution in biomedical full-text articles with domain dependent features. In *Computer Technology and Development (ICCTD), 2010 2nd International Conference on*, pages 616–620. IEEE.

Prateek Jindal and Dan Roth. 2013. Using domain knowledge and domain-inspired discourse model for coreference resolution for clinical narratives. *Journal of the American Medical Informatics Association*, 20(2):356–362.

Halil Kilicoglu, Graciela Rosemblat, Marcelo Fiszman, and Thomas C Rindflesch. 2016. Sortal anaphora resolution to enhance relation extraction from biomedical literature. *BMC bioinformatics*, 17(1):1.

Jin-Dong Kim, Tomoko Ohta, and Jun'ichi Tsujii. 2008. Corpus annotation for mining biomedical events from literature. *BMC bioinformatics*, 9(1):1.

Jin-Dong Kim, Tomoko Ohta, Sampo Pyysalo, Yoshinobu Kano, and Jun'ichi Tsujii. 2009. Overview of bionlp'09 shared task on event extraction. In *Proceedings of the Workshop on Current Trends in Biomedical Natural Language Processing: Shared Task*, pages 1–9. Association for Computational Linguistics.

Thomas Lumley. 2002. Network meta-analysis for indirect treatment comparisons. *Statistics in medicine*, 21(16):2313–2324.

Vincent Ng and Claire Cardie. 2002a. Identifying anaphoric and non-anaphoric noun phrases to improve coreference resolution. In *Proceedings of the 19th international conference on Computational linguistics-Volume 1*, pages 1–7. Association for Computational Linguistics.

Vincent Ng and Claire Cardie. 2002b. Improving machine learning approaches to coreference resolution. In *Proceedings of the 40th Annual Meeting on Association for Computational Linguistics*, pages 104–111. Association for Computational Linguistics.

F. Pedregosa, G. Varoquaux, A. Gramfort, V. Michel, B. Thirion, O. Grisel, M. Blondel, P. Prettenhofer, R. Weiss, V. Dubourg, J. Vanderplas, A. Passos, D. Cournapeau, M. Brucher, M. Perrot, and E. Duchesnay. 2011. Scikit-learn: Machine learning in Python. *Journal of Machine Learning Research*, 12:2825–2830.

Haoruo Peng, Daniel Khashabi, and Dan Roth. 2015. Solving hard coreference problems. *Urbana*, 51:61801.

Michael Poprat, Elena Beisswanger, and Udo Hahn. 2008. Building a biowordnet by using wordnet's data formats and wordnet's software infrastructure: a failure story. In *Software engineering, testing, and quality assurance for natural language processing*, pages 31–39. Association for Computational Linguistics.

Sameer Pradhan, Noemie Elhadad, Brett R South, David Martinez, Amy Vogel, Hanna Suominen, Wendy W Chapman, and Guergana Savova. 2013. Task 1: Share/clef ehealth evaluation lab.

Preethi Raghavan, Eric Fosler-Lussier, and Albert M Lai. 2012. Exploring semi-supervised coreference resolution of medical concepts using semantic and temporal features. In *Proceedings of the 2012 Conference of the North American Chapter of the Association for Computational Linguistics: Human Language Technologies*, pages 731–741. Association for Computational Linguistics.

Preethi Raghavan, Eric Fosler-Lussier, Noémie Elhadad, and Albert M Lai. 2014. Cross-narrative temporal ordering of medical events. In *ACL (1)*, pages 998–1008. Citeseer.

Karthik Raghunathan, Heeyoung Lee, Sudarshan Rangarajan, Nathanael Chambers, Mihai Surdeanu, Dan Jurafsky, and Christopher Manning. 2010. A multipass sieve for coreference resolution. In *Proceedings of the 2010 Conference on Empirical Methods in Natural Language Processing*, pages 492–501. Association for Computational Linguistics.

Wee Meng Soon, Hwee Tou Ng, and Daniel Chung Yong Lim. 2001. A machine learning approach to coreference resolution of noun phrases. *Computational linguistics*, 27(4):521–544.

Rodney L Summerscales, Shlomo Argamon, Shangda Bai, Jordan Huperff, and Alan Schwartz. 2011. Automatic summarization of results from clinical trials. In *Bioinformatics and Biomedicine (BIBM), 2011 IEEE International Conference on*, pages 372–377. IEEE.

Rodney L Summerscales. 2013. *Automatic summarization of clinical abstracts for evidence-based medicine*. Ph.D. thesis, Illinois Institute of Technology.

Antonio Trenta, Anthony Hunter, and Sebastian Riedel. 2015. Extraction of evidence tables from abstracts of randomized clinical trials using a maximum entropy classifier and global constraints. *arXiv preprint arXiv:1509.05209*.

Ozlem Uzuner, Andreea Bodnari, Shuying Shen, Tyler Forbush, John Pestian, and Brett R South. 2012. Evaluating the state of the art in coreference resolution for electronic medical records. *Journal of the American Medical Informatics Association*, 19(5):786–791.

Byron C Wallace, Issa J Dahabreh, Christopher H Schmid, Joseph Lau, and Thomas A Trikalinos. 2013. Modernizing the systematic review process to inform comparative effectiveness: tools and methods. *Journal of comparative effectiveness research*, 2(3):273–282.

Byron C Wallace, Jol Kuiper, Aakash Sharma, Mingxi (Brian) Zhu, and Iain J. Marshall. 2016. Extracting pico sentences from clinical trial reports using supervised distant supervision. *Journal of Machine Learning Research (JMLR)*.

Xiaofeng Yang, Jian Su, Guodong Zhou, and Chew Lim Tan. 2004. An np-cluster based approach to coreference resolution. In *Proceedings of the 20th international conference on Computational Linguistics*, page 226. Association for Computational Linguistics.

Bishan Yang, Claire Cardie, and Peter Frazier. 2015. A hierarchical distance-dependent bayesian model for event coreference resolution. *arXiv preprint arXiv:1504.05929*.

Katsumasa Yoshikawa, Sebastian Riedel, Tsutomu Hirao, Masayuki Asahara, and Yuji Matsumoto. 2011. Coreference based event-argument relation extraction on biomedical text. *Journal of Biomedical Semantics*, 2(5):1.

Chun-Nam John Yu and Thorsten Joachims. 2009. Learning structural svms with latent variables. In *Proceedings of the 26th Annual International Conference on Machine Learning*, ICML '09, pages 1169–1176, New York, NY, USA. ACM.

Jin Zhao, Praveen Bysani, and Min-Yen Kan. 2012. Exploiting classification correlations for the extraction of evidence-based practice information. In *AMIA*.

Hybrid methods for ICD-10 coding of death certificates

Pierre Zweigenbaum
LIMSI, CNRS,
Université Paris-Saclay,
F-91405 Orsay, France
`pz@limsi.fr`

Thomas Lavergne
LIMSI, CNRS, Univ. Paris-Sud
Université Paris-Saclay,
F-91405 Orsay, France
`lavergne@limsi.fr`

Abstract

ICD-10 coding of death certificates has received renewed attention recently with the organization of the CLEF eHealth 2016 clinical information extraction task (CLEF eHealth 2016 Task 2). This task has been addressed either with dictionary projection methods or with supervised machine learning methods, but none of the participants have tried to design hybrid methods to process these data. The goal of the present paper is to explore such hybrid methods. It proposes several hybrid methods which outperform both plain dictionary projection and supervised machine learning on the training set. On the official test set, it obtains an F-measure of 0.8586 which is 1pt above the best published results so far on this corpus ($p < 10^{-4}$). Moreover, it does so with no manual dictionary tuning, and thus has potential for generalization to other languages with little effort.

1 Introduction

Biomedical information processing crucially relies on a normalized representation of medical information in the form or standardized terminologies and ontologies, be it for clinical care (SNOMED, LOINC), for public health statistics and health management (International Classification of Diseases) or for literature search (MeSH). Automatically generating such a normalized representation from naturally occurring sources such as text is therefore a long-studied goal (Wingert et al., 1989). Basically, it consists in deciding which concepts in the target representation (e.g., signs and symptom concepts in SNOMED CT, or disease classes in the ICD-10 classification) best represent the contents of a given text (e.g., a patient discharge summary). It can

be decomposed into the detection of text mentions of biomedical concepts of the suitable types (entity recognition) and the determination of the target concepts (concept normalization) which best represent the text mentions in the context of the source text and the given use case. The state of the art of biomedical entity recognition and biomedical concept normalization has been established and published in a number of shared tasks which addressed clinical texts (Pestian et al., 2007; Uzuner et al., 2007; Uzuner et al., 2011; Suominen et al., 2013), biomedical literature (Kim et al., 2011; Nédellec et al., 2015), sometimes in multiple languages (Suominen et al., 2013; Névéol et al., 2016).

This paper focuses on ICD-10 coding. ICD coding has been studied in the past (e.g., as early as (Wingert et al., 1989)), but only recently has a large dataset been released for ICD-10 coding of death certificates (Névéol et al., 2016). In that context, Névéol et al. (2016) mention that participants in the CLEF eHealth 2016 ICD-10 coding task either used dictionary-based methods or supervised machine learning methods, and that none tried hybrid methods. The goal of this paper is to explore this direction. Our contributions are the following:

- We explore hybrid methods for ICD-10 coding which combine dictionary projection and supervised machine learning.

- We show that simple hybrid combinations with union and intersection yield improved results.

- We propose methods which improve the precision of dictionary projection, including hybrid 'calibration' methods.

- The methods which fare best on the training corpus, when applied to the test corpus, are

96

Proceedings of the Seventh International Workshop on Health Text Mining and Information Analysis (LOUHI), pages 96–105,
Austin, TX, November 5, 2016. ©2016 Association for Computational Linguistics

on par with the best published results on this corpus, with no manual dictionary tuning, and have thus potential for generalization to other languages with little effort.

In the remainder of the paper, we report the methods used by the best-performing participants in the CLEF eHealth 2016 shared task (Section 2), present the methods we explored and the data on which we applied them (Section 3), the results we obtained on the development and test data (Section 4), discuss them (Section 5) and conclude (Section 6).

2 Related Work

When producing normalized concepts from medical texts, most methods use dictionary-based lexical matching or supervised machine-learning. Most dictionary-based methods use the UMLS (Bodenreider, 2004) or one of its included vocabularies, such as the ICD-10 classification. MetaMap (Aronson and Lang, 2010) is the most used system for English: it takes advantage of the term variants present in the UMLS MetaThesaurus and of the morphological knowledge provided by the UMLS Specialist Lexicon. Knowledge-lean methods based on approximate dictionary look-up have also been proposed (Zhou et al., 2006).

Some studies have addressed the ICD-10 coding of death certificates. Koopman et al. (2015a) classified Australian death certificates into 3-digit ICD-10 codes such as *E10* with SVM classifiers based on n-grams and SNOMED CT concepts, and with rules. They also trained SVM classifiers (Koopman et al., 2015b) to find ICD-10 diagnostic codes for death certificates. In contrast to the above-mentioned CLEF eHealth shared task, they only addressed cancer-related certificates: they set-up a first-level classifier to detect cases of cancer then a second-level classifier to refine it into a specific type. Another difference from CLEF eHealth is that they remained at the level of 3-digit ICD-10 codes (e.g., *C00*, *C97*) instead of the full 4-digit level usually required for ICD-10 coding (e.g., *C90.9*). Another important difference is that they targeted the underlying cause of death, i.e., one diagnosis per death certificate, whereas the CLEF eHealth task requires to determine all the diagnoses mentioned in each statement of a given death certificate.

Two ICD coding shared task were organized so far. The Computational Medicine Center (CMC) challenge (Pestian et al., 2007) targeted ICD-9-CM disease coding from outpatient chest x-ray and renal procedures, whose clinical history and impression sections provide most support for coding. The dataset contained 978 documents for training and 976 documents for testing. It targeted a small subset of 45 ICD-9-CM codes, designed in such a way that every one of the 94 distinct combination of codes present in the test set were seen in the training set. The best system was based on a supervised classifier (a Decision Tree) and obtained an F-measure of 0.89 on the test set.

The CLEF eHealth 2016 ICD-10 coding task (Névéol et al., 2016) provided a dataset which consisted of death certificates in French. These death certificates were provided by CépiDc, the WHO collaborating center which manages ICD-10 coding of death certificates in France. We reproduce the corpus statistics from the task organizers' paper in Table 1. The task was defined at the level of each statement (line) in a death certificate: one statement could be associated with 0, 1 or more ICD-10 codes which represent causes of death at various levels in the causal chain which led to the death. Statements have a length which varies from 1 to 30 words, with outliers at 120 words and the most frequent length at 2 tokens. They are thus much shorter than the CMC challenge texts.

	Training (2006–2012)	Test (2013)
Documents	65,844	27,850
Lines	195,204	80,899
Tokens	1,176,994	496,649
Total ICD codes	266,808	110,869
Unique ICD codes	3,233	2,363

Table 1: The CépiDC French Death Certificates Corpus (from Névéol et al.).

The full dataset contained death certificates from 2006 to 2013. In a natural use case, death certificates of former years have already been coded and are available as examples to code new death certificates. Therefore the test corpus contained certificates of year 2013, whereas the training corpus contained certificates of years 2006–2012. There was therefore no guarantee that a code needed in 2013 had been used in 2006-2012: a posteriori analysis reveals that 224 of the 2,363 unique codes used in 2013 were not used in 2006–2012. Besides, as can be seen in Table 1, the size of the corpus is much larger than that of the CCMC challenge, as well as the number of target codes.

Table 2 shows examples statements from the dataset; we provided English translations for the reader's convenience.

Statement + English gloss	Codes
surinfection superinfection	B99
insuffisance respiratoire aiguë acute respiratory failure	J960
arrêt cardio-respiratoire hypoxémique hypoxaemic cardio-respiratory arrest	R092, R090
Hypertrophie ventriculaire gauche concentrique d'étiologie indéterminée Concentric left ventricular hypertrophy of unknown origin	I517
Epilepsie séquellaire à AVC sylvien droit, AC/FA chronique, insuffisance cardiaque congestive, insuffisance rénale, atélectasie pulmonaire Sequelar epilepsy with right sylvian stroke, chronic atrial fibrillation/cardiac arrhythmia, congestive heart failure, renal failure, pulmonary atelectasis	J981, I500, G409, I48, I64, N19

Table 2: Statement examples with their associated ICD-10 codes, with English glosses. Code order does not necessarily align with text order.

CLEF participants were also provided with dictionaries created by CépiDc for their own use. Each dictionary included (term, ICD-10 code, related code 1, related code 2) quadruplets. We did not use the two 'related codes', hence only consider (term, ICD-10 code) pairs in the remainder of this paper. Four dictionaries were provided: one used over the years 2006–2010 (157,001 lines), one for 2011 (156,937 lines), one for 2012 (158,163 lines), and one for 2013 (144,905 lines). These dictionaries reflect changes in coding practice over the years, either caused by changes in international ICD contents or coding rules, or by newly encountered expressions which were not covered in previous years, or by improvements in CépiDc's dictionary management.

The top two systems at the CLEF eHealth ICD-10 coding task used two different methods.

Van Mulligen et al. (2016) relied on ICD dictionaries built from the shared task data. Their baseline dictionary used the term-code associations seen in the shared task training set, and their expanded dictionary also used the above-mentioned CépiDc dictionaries. Various filters were applied to these dictionaries, based on the ambiguity of the term-code associations. Their dictionary projection method used the Solr information-retrieval system to cope with the large number of entries in the lexicon efficiently. After measuring its performance on the training corpus, they post-processed their system output to block term-code associations with a precision on the training set lower than a given threshold selected by optimizing F-measure on the training set. They obtained the top precision, recall, and F-measure published so far on this dataset: P=0.886, R=0.813, F=0.848 in their top run using the expanded dictionary, or P=0.890, R=0.803, F=0.844 in their second run using the baseline dictionary.

Instead of trying to spot occurrences of known terms or variants in the input statements and then normalize them to ICD codes, Dermouche et al. (2016) addressed the task as a text classification problem: given a short text, compute a class, here an ICD-10 code. They used a supervised machine learning method (SVM) with bags of words after text preprocessing. They also tested transformations of the obtained vector space representation with topic models. The precision of their best submitted run (P=0.882) was very close to the that of the top system but their recall and F-measure were lower (P=0.882, R=0.655, F=0.752). The probable reason for their lower recall was that they produced one code per statement (mono-label classification), whereas given the data in Table 1, we can compute that there was an average of 1.37 codes per source statement both in the training corpus and in the test corpus. If a similar method could address multi-label classification and scale its recall linearly, it would reach a recall of $0.655 \times 1.37 = 0.897$, even higher than the dictionary projection method, which naturally performs multi-label classification.

As mentioned in the introduction, Névéol et al. (2016) observed that no participant in the CLEF eHealth 2016 ICD-10 coding task tried hybrid methods which would combine dictionary projection and supervised machine learning. Exploring this direction is the goal of this paper.

3 Methods

We set up a simple dictionary projection method and a supervised machine learning method, then designed hybrid methods based on one or both of them.

We first processed each statement as follows: conversion to lower case, tokenization (with an NLTK regular expression), stop word removal (French NLTK); diacritic removal (Unicode 'NFD' normalization), correction of some spelling errors based on the words present in the training corpus and in the CépiDc dictionaries, stemming (Snowball French stemmer).

3.1 Dictionary projection

Dictionary projection relies on the expressions present in a dictionary to spot mentions of concepts in a text. We pre-processed the CépiDc dictionaries in the same way as the death certificate statements: as a result, each dictionary entry links a sequence of normalized tokens to one or more ICD codes. For term matching efficiency, each dictionary was stored as a Trie. Given a dictionary, an input sequence of tokens is processed as follows. The input sequence of tokens is scanned for the first match. In case of multiple matches, the longest match is retained. After a match, scanning resumes right after the end of the match. The output of the process is a (possibly empty) list of matched dictionary entries together with their positions in the input sequence.

No processing of negations was performed because statements are very short and negations are infrequent. For instance, only 82 occurrences of the negation *pas* (*no/not*) were found in the training corpus (i.e., in 0.04% of the statements), and 240 occurrences of the negation *sans* (*without*) (0.12%).

A dictionary entry may lead to $0:n$ codes. Depending on how the dictionary was built, the same code may have been recorded multiple times: this number of times is recorded in the dictionary. We have tested the following selection strategies in case of multiple outputs for a given entry:

all All codes are returned.

best The most frequently recorded code is returned. In case of a tie, a random choice is performed.

boiu (Best Only If Unambiguous): The most frequently recorded code is returned only if there is no tie, else no result is returned.

Dictionary projection can use any available dictionary which links terms to ICD codes. Here we tested only those provided by CépiDc to the CLEF eHealth participants: the use of other dictionaries which could be built for instance from the training corpus, from the ICD-10 terms themselves, or from the UMLS, is left for future work.

3.2 Supervised classification

Supervised classification is not the focus of this paper, therefore we only present here our best current model. It uses a linear SVM classifier and the following method and features:

- Linear SVM (scikit-learn's LinearSVC with default parameters, which relies on liblinear)

- Tokens (t), obtained after the above-mentioned pre-processing step. We also tested token n-grams up to 5, but this did not improve the results.

- Character trigrams ($c3$): spelling errors are frequent in the certificates; representing a statement by its overlapping character trigrams provides a degree of robustness to spelling errors.

- Coding Year (y): coding rules change over the years, and the same statement seen at two different dates may be coded differently because of such changes. Therefore we found it useful to include 2×9 features instantiated for $y \in [2006 \ldots 2014]$: '> y' or '$\leq y$' depending on the value of the Coding Year (e.g., a statement of 2011 will have '>2006', ... '>2010', '$\leq$2011', ... '$\leq$2014'.

This supervised classifier uses no information on ICD terms or codes other than that present in its training corpus.

3.3 Union and intersection of classifiers

The union of the outputs of two classifiers is a very simple method to combine them. It is useful when the individual classifiers lack recall, and preferably have a high enough precision. The ideal situation occurs when individual classifiers output different correct predictions (in which case the resulting recall will be higher than the best recall of the individual classifiers) and when the individual classifiers make errors on the same inputs (in which case the resulting number of false positives will be lower than the sum of the individual false positives).

Conversely, the intersection of two classifiers is a possible method to increase their precision. A high-precision classifier is useful for pre-annotation. In the actual coding process at CépiDc, human coders spend a sizable part of their time assigning codes which are easy to determine. Pre-annotating these codes with a reliable, high-precision system before presenting death certificates to human coders would enable them to browse through these pre-assigned codes quickly. This would save human coding time which could be reassigned to solving more difficult cases.

3.4 Calibration

A prediction method, for instance dictionary projection, can be 'calibrated' by training a classifier to detect its errors. Calibration takes into account the distribution of codes and of prediction success in the training split, thereby adding data-driven knowledge to the application of the expert-produced dictionary. It automatically spots the main deficiencies of the dictionary projection and blocks them. In this respect, it is closely related to the error analysis process which a human expert performs when applying their dictionary to a new dataset: error spotting, then correction. In the human process, correction can take the form of simple post-processing rules which filter out output codes known to be often erroneous. It can also come from data-driven tuning of the dictionary by measuring the performance of its entries on the training corpus and selecting an appropriate threshold to prune low-performance entries, as in (Van Mulligen et al., 2016). This is exactly what is performed automatically by the classifier we train.

We trained a classifier with the following conditions:

- Classifier: Linear SVM (scikit-learn's LinearSVC with default parameters).

- Features: individual code predicted by the CépiDc dictionary (see below Section 3.5), prefixed by *code:* (e.g., *code:R068*); we also tested the addition of the statement tokens (obtained by the same process as described above).

- Classes: True (meaning the predicted code is correct) / False (meaning it is incorrect).

- Training: our training split (see below: 185k statements) for development, the full training corpus for testing.

When testing, the trained classifier was applied to each individual code predicted by the dictionary projection. If the classifier's output was the False class, the predicted code was removed from the dictionary projection output.

3.5 Data

We used the CépiDc data provided by the CLEF eHealth 2016 clinical information extraction task (CLEF eHealth 2016 Task 2) to the challenge participants (Névéol et al., 2016). The statistics of the training and test corpora are described in Section 2. To emulate the test conditions in our development phase, we also split the training corpus based on the dates of the certificates: the last 10,000 statements (1141 unique codes) made up our test split, while the first 185,204 statements (13,300 codes, 3,200 unique) constituted our training split. Only 11 codes were present in the test split but absent from the training split.

Python 3.5.2 was used for the programs, with scikit-learn 0.17.1, within Anaconda 4.0.0.

3.6 Experimental protocol and evaluation

Teams were allowed to submit up to three runs to the task. In the present work, we emulated the same situation and selected three methods to run on the test corpus based on their F-measures in our experiments on the training corpus. This prevented us from biasing the final results by tuning them on the test corpus. To apply these methods to the test corpus, we retrained them on the full training corpus with a more recent dictionary:

- The supervised classifier (Linear SVM, *tc3y*) was trained on the full training corpus.

- Dictionary projection methods used the 2012 dictionary instead of the 2011 dictionary.

- Dictionary projection was calibrated on the full training corpus.

- Supervised classifier and calibrated dictionary projection were applied to the test corpus.

- The union of their results was computed and used as final predictions.

Precision, recall and F-measure were computed for each experiment, by our own programs for convenience during development; when applied to the test corpus, they were computed with the official scoring program provided to the CLEF eHealth participants.

4 Results

4.1 Development: results on the test split of the training corpus

The SVM classifier with tokens, character trigrams, and year coded (henceforth *tc3y*), was trained on our training split and applied to our test split, on which it obtained P=0.9010, R=0.6774, and F=0.7734.

We tested the four dictionaries and our three dictionary output selection methods on our test split. Table 3 shows that the 2011 dictionary obtains the best precision, recall and F-measure, closely followed by the 2012 dictionary. As could be expected, the *all* method always produced the highest recall, whereas the *boiu* method always produced the highest precision. *boiu* also obtained the highest F-measure. The top F-measure was thus obtained with the 2011 dictionary and *boiu*, at P=0.8048, R=0.6475 and F=0.7176. We therefore retained the 2011 dictionary for further experiments on our test split (2012 data). We also assumed that following the same pattern for the official test data, dated in 2013, the 2012 dictionary should be most suitable. As a safety check, we tested the 2012 dictionary on our test split in the same conditions as the

2011 dictionary, and observed that it obtained similar results—slightly inferior, by a maximum of 0.1 pt P, R or F.

Dict	Sel	# Sys	TP	P	R	F
2006	boiu	10720	8470	0.7901	0.6368	0.7052
2006	best	12977	9117	0.7026	0.6855	0.6939
2006	all	18458	10133	0.5490	0.7619	0.6381
2011	boiu	10701	8612	0.8048	0.6475	0.7176
2011	best	12978	9335	0.7193	0.7019	0.7105
2011	all	18722	10491	0.5604	**0.7888**	0.6552
2012	boiu	10580	8485	**0.8020**	0.6380	**0.7106**
2012	best	12970	9276	0.7152	0.6974	0.7062
2012	all	18520	10469	0.5653	**0.7871**	0.6580
2013	boiu	10550	8106	0.7683	0.6095	0.6797
2013	best	13095	8951	0.6835	0.6730	0.6782
2013	all	19285	9956	0.5163	0.7486	0.6111

Table 3: Dictionary experiments on our test split: CépiDc dictionaries (Dict), 10,000 statements, 13,300 codes: all statements date from year 2012. Sel = Selection method: boiu = best only if unambiguous, best = most frequent code (random choice in case of tie), all = all codes. # Sys = number of system-predicted codes. TP = true positives. P = precision, R = recall, F = F-measure.

Table 4 shows the 2011 dictionary results without (–) and with (*c, c-t*) calibration. Calibration based only on the dictionary-proposed code (*Cal=c*) boosts precision by 12pt (*boiu*) to 33pt (*all*) and F-measure by 2.6pt (*boiu*) to 14pt (*all*), while only reducing recall by 2.5pt (*boiu*) to 6pt (*all*). Additionally taking into account the tokens of the coded statement in calibration (*Cal=c-t*) adds another 1.7pt (*boiu* or *all*) to 1.9pt (*best*) to precision and 0.25pt (*boiu*) to 0.6pt (*all*) to F-measure, with a decrease of recall by 0.15pt (*all*) to 0.4pt (*boiu* or *best*). Altogether, calibration is therefore highly efficient on our test split to increase precision and F-measure. The highest precision is obtained with *boiu, c-t* while the highest F-measure is obtained with *all, c-t*.

We performed the union and the intersection of the outputs of the SVM supervised classifier and of the dictionary projection. The results are reported in Table 5.

Union with the non-calibrated dictionary projection decreased its precision only by 1pt (*boiu*) or even increased it by 1 or 2pt (*best, all*) because the supervised classifier had a much higher precision, at

Sel	Cal	# Sys	TP	P	R	F
boiu	–	10701	8612	0.8048	**0.6475**	0.7176
boiu	c	8971	8276	0.9225	0.6223	0.7432
boiu	c-t	8749	8221	0.9397	0.6181	**0.7457**
best	–	12978	9335	0.7193	**0.7019**	0.7105
best	c	9769	8823	0.9032	0.6634	0.7649
best	c-t	9514	8773	**0.9221**	0.6596	**0.7691**
all	–	18722	10491	0.5604	0.7888	0.6552
all	c	10809	9631	0.8910	0.7241	0.7990
all	c-t	10585	9610	**0.9079**	0.7226	0.8047

Table 4: 2011 dictionary calibration experiments on our test split. Cal = calibration: – (none), c (dictionary code), t (source tokens).

Sel	Cal	# Sys	TP	P	R	F
svm (linear)	tc3y	9010		0.9010	0.6774	0.7734
Union						
boiu	–	14188	11303	0.7967	0.8498	0.8224
boiu	c	12670	11153	0.8803	0.8386	0.8589
boiu	c-t	12447	11087	0.8907	0.8336	0.8612
best	–	15894	11566	0.7277	0.8696	0.7924
best	c	13017	11313	0.8691	0.8506	0.8597
best	c-t	12719	11224	0.8825	0.8439	0.8628
all	–	20836	12034	0.5776	0.9048	0.7051
all	c	13414	11519	0.8587	0.8661	0.8624
all	c-t	13142	11457	0.8718	0.8614	0.8666
Intersection						
boiu	–	6293	6128	0.9738	0.4608	0.6255
boiu	c	6291	6127	0.9739	0.4607	0.6255
boiu	c-t	6302	6144	0.9749	0.4620	0.6269
best	–	7084	6779	0.9569	0.5097	0.6651
best	c	6752	6520	0.9656	0.4902	0.6503
best	c-t	6795	6559	0.9653	0.4932	0.6528
all	–	7886	7467	0.9469	**0.5614**	**0.7049**
all	c	7395	7122	0.9631	0.5355	0.6883
all	c-t	7443	7163	0.9624	0.5386	0.6906

Table 5: 2011 dictionary experiments on our test split: Union and intersection of Linear SVM and dictionary results.

the same time boosting recall by 12 to 20pt, reaching a maximum of 0.9048. Union with the calibrated dictionary projection decreased its precision by at most 5pt (*boiu, c-t*), maintaining a very reasonable P=0.86–0.89. Recall was boosted by 14 to 19pt, leading to a record F-measure of 0.8666.

Again, *all* obtained the highest recall and also the highest F-measure, achieving records of R=0.8661 (–) and F=0.8666 (*c-t*, both with quite balanced P, R, F. The *all c-t* combination was thus a natural candidate to run on the official test corpus.

With intersection, the obtained precision gained 3.5pt over the best so far, reaching 0.96–0.97, while losing 16–19pt of recall at 0.46–0.54 compared to the calibrated dictionary projection. Here again, *boiu* obtained the highest precisions with the top at 0.9749 (*c-t*). Intersection yields a 58% reduction of the best error rate so far from 6% to 2.5%. With such a low error rate, pre-annotation becomes viable and would cater for not far from one half of the number of codes to produce (R=0.4620). For information, we added this precision-oriented configuration (*boiu c-t*) to the three F-measure-oriented configurations to be run on the official test corpus.

4.2 Results on the test corpus

The best F-measure on the training corpus was obtained by the union of the SVM classifier and the *all* dictionary projection calibrated with token features (*all-c-t*), therefore we selected this method as our Run 1. We wanted to diversify our tests, therefore also selected two more precise runs: (*ii*) the union of the SVM classifier and the *boiu* dictionary projection calibrated with token features (*boiu-c-t*), and (*iii*) the union of the SVM classifier and the *boiu* dictionary projection calibrated with no extra features (*boiu-c*). We applied these methods to the test corpus in the manner presented above.

The results obtained on the official test corpus are very close to those on our test split of the training corpus: there is a constant difference of only –0.8pt in F-measure, and a similarly small decrease of less than 1pt in precision and recall for the three runs. This shows that the tested methods do not overfit the training corpus. As a consequence, the order of results on the test corpus reproduces that of the test split: highest F-measure and recall for *u(lsvcd(tc3y),d2012-all-c-t)*, highest precision for *u(lsvcd(tc3y),d2012-boiu-c-t)*.

The F-measures of the three selected runs exceed that of the best CLEF eHealth participant (P=0.886, R=0.813, F=0.848) by 0.3 to 1pt and their recalls do so by 1 to 4pt, whereas the precisions of these runs are below the best CLEF precision (P=0.890) by 0.6 to 2.5pt. Because of the large size of the test corpus, all of the differences from the best CLEF run (see Table 6) are significant (tested with ap-

Method	P	R	F
svm (tc3y)	0.8938	0.6645	0.7623
u(svm,d-all-c-t)	0.8656^{-4}	$\mathbf{0.8517}^{-4}$	$\mathbf{0.8586}^{-4}$
u(svm,d-boiu-c-t)	$\mathbf{0.8840}$	0.8242^{-4}	0.8531^{-4}
u(svm,d-boiu-c)	0.8751^{-4}	0.8282^{-4}	0.8510^{-3}
i(svm,d-boiu-c-t)	$\mathbf{0.9703}$	0.4500	0.6148

Table 6: Tests on the official test corpus. Evaluation with the official program. u(a,b) = union(a,b). i(a,b) = intersection(a,b). svm is a linear SVM with features tc3y. d- = dictionary (2012). In Union results, superscripts represent the power of the p-value of significance testing for the difference with the best published result so far (Van Mulligen et al., 2016): $-3 = p < 10^{-3}$, $-4 = p < 10^{-4}$. Note that the values of P are higher in (Van Mulligen et al., 2016) (the difference is significant in two cases out of three) whereas the values of R and F are better in the present Union results (differences are always significant).

proximate randomization with 10,000 permutations, $p = 10^{-4}$ for all except $p = 0.6 \times 10^{-3}$ for the difference of 0.3pt in F-measure), except the difference of 0.2pt in precision ($p = 0.104$). Note however that the methods and experiments presented in the present paper benefited from extra time invested after the official CLEF eHealth run submissions, so that a comparison with results obtained during the shared task time frame does not reflect differences in quality of the involved teams.

5 Discussion

5.1 Calibration

Calibration proved highly efficient in the present setting.

For instance, calibration of *boiu* output with only code-based classification (*boiu c* in Table 4) filters out 258 instances of ICD-10 code *C809,* Malignant neoplasms of ill-defined, secondary and unspecified sites which dictionary projection assigned to our test split, among which only 15 were true positives and 243 were false positives. The dictionary happens to have 509 entries for this code, among which the single word *cancer*. Because of the longest match strategy, this entry generally does not fire because longer entries including this word exist and will match instead. However, it acts as a default entry which may be used in inappropriate contexts.

Because we applied calibration to filter out some

target codes, it blocks full sets of dictionary entries (for instance, the 509 entries for *C809*). A finer-grained method might try to filter out specific entries instead, and maybe still obtain a good increase in precision while limiting the associated loss in recall.

5.2 Union and intersection

Union and intersection are very simple combination methods. They played their expected roles in our experiments. Because we started from predicted results with precisions above 0.90, union was able to keep a high enough precision (up to 0.89 on the training set and 0.88 on the test set). The fact that it also led to a strongly increased recall shows that dictionary projection and our mono-label supervised classifier produced complementary results.

Given that we started from high-precision results, intersection was interesting to obtain very-high-precision classifiers. On the training set, the obtained precision ranged from 0.94 to 0.97, with associated recalls decreasing from 0.56 to 0.46. A study of the 3% resisting codes is left for future work. The highest-precision configuration, when applied to the test set, also reached a 0.97 precision with a 0.45 recall. This means that nearly one half of the test statements can be annotated automatically with an error rate of only 3%. This makes pre-annotation of death certificates with these methods a viable proposal to save human coding time.

5.3 Dictionary projection as a classifier feature

A very simple way to combine two classifiers is to use the output of one of them as a feature for the other. As suggested by an anonymous reviewer, we tested this scheme by using the ICD codes detected by dictionary projection (with the *boiu, best,* or *all* selection method) as an additional feature for the supervised classifier. We trained and tested the SVM supervised classifier with this additional feature based on the 2011 dictionary. This improved P, R and F by about 1pt on our test split (P=0.9154, R=0.6883, F=0.7858 with the *best* selection method). We then computed the intersection of the obtained classifier with the dictionary results (with and without calibration), as performed before to obtain the results in Table 5. This increased the best union F-measure (*all, all-c-t*) by up to 0.3pt at 0.8697 (with *all* selection method) as well as all other union F-measures, but obtained a lower best intersection precision (–0.4pt at 0.9711, *boiu, boiu-c-t*). The influence of the selection method used in dictionary projection for feature creation was minor. This additional combination might increase again the F-measure on the test corpus, but was not tested in this paper.

5.4 Generalizability

The ICD-10 coding of death certificates is a process which is performed world-wide in a variety of languages. Efforts have been spent in various countries to develop dictionaries such as that of the CépiDc in France. An important feature of the methods we have presented here is that they are readily portable to other languages. The only language-dependent parts of our methods are diacritic removal (which generalizes to all Unicode languages to which the 'NFD' normalization applies), stemming (which is readily available for dozens of languages), some offline spelling correction (which generalizes to many alphabetic languages), and the use of character trigrams (which generalizes to alphabetic languages). No manual dictionary entry development or tuning was performed at all. Moreover, the supervised method already yields a high precision even without any dictionary at all, provided a sufficient number of training examples are available.

Therefore our methods and system should be applicable with no or little effort to a number of other languages.

6 Conclusion

We explored hybrid methods which combine simple dictionary projection and mono-label supervised classification. Our starting point was a dictionary projection method which obtained a higher recall and a supervised classification method which obtained a higher precision. Calibration strongly improved the precision of dictionary projection, making it higher than that of the supervised classifier. Union of calibrated dictionary projection results and supervised classification results improved the recall of both of them while keeping a high enough precision, leading to the highest F-measure on the training corpus. Intersection of calibrated dictionary projection results and supervised classification results

obtained a record precision of 0.97 while producing codes for a little less than one half of the statements. This is a suitable configuration for automatic pre-annotation of death certificates which could save time to human coders. These experiments were performed on the training corpus: when applying the best development configurations to the test corpus, they led to three runs (F=0.8510–0.8586) which are all above the best published F-measure so far (F=0.848, significant at $p < 10^{-4}$) on this dataset. An important advantage of these methods is that they only relied on the data provided by the French coding center, CépiDc: if similar organizations in other countries have similar data, these methods should be readily applicable with little change to these new data.

In future work we plan to improve the individual methods and test more hybrid methods. Using more complete dictionaries is a way to improve the recall and maybe precision too of dictionary projection. Changing the supervised classification to perform multi-label classification is a direction to improve the recall of the supervised classifier. Calibrating the dictionary at the level of individual entries instead of target codes might also limit the loss of dictionary projection recall when increasing its precision.

Acknowledgments

We thank the CLEF eHealth challenge organizers for providing the data used in the present work and Jan Kors (ERASMUS team) for giving us access to their results for significance testing. This work was partially funded by the European Union's Horizon 2020 Marie Sklodowska Curie Innovative Training Networks—European Joint doctorate (ITN-EJD) Under grant agreement No:676207 (MiRoR). Finally, we thank the anonymous reviewers for their very relevant comments which helped improve the paper.

References

Alan R Aronson and François-Michel Lang. 2010. An overview of MetaMap: historical perspective and recent advances. *J Am Med Inform Assoc*, 17(3):229–36.

Olivier Bodenreider. 2004. The Unified Medical Language System (UMLS): Integrating biomedical terminology. *Nucleic Acids Research*, 32(Database issue):D267–270.

Mohammed Dermouche, V Looten, Rémy Flicoteaux, Sylvie Chevret, J Velcin, and Namik Taright. 2016. ECSTRA-INSERM @ CLEF eHealth2016-task 2: ICD10 code extraction from death certificates. In *CLEF 2016 Online Working Notes*. CEUR-WS.

Jin-Dong Kim, Sampo Pyysalo, Tomoko Ohta, Robert Bossy, Ngan Nguyen, and Jun'ichi Tsujii. 2011. Overview of BioNLP shared task 2011. In *Proceedings of BioNLP Shared Task 2011 Workshop*, pages 1–6, Portland, Oregon, USA, June. Association for Computational Linguistics.

Bevan Koopman, Sarvnaz Karimi, Anthony Nguyen, Rhydwyn McGuire, David Muscatello, Madonna Kemp, Donna Truran, Ming Zhang, and Sarah Thackway. 2015a. Automatic classification of diseases from free-text death certificates for real-time surveillance. *BMC Med Inform Decis Mak*, 15:53.

Bevan Koopman, Guido Zuccon, Anthony Nguyen, Anton Bergheim, and Narelle Grayson. 2015b. Automatic ICD-10 classification of cancers from free-text death certificates. *Int J Med Inform*, 84(11):956–965, November.

Claire Nédellec, Jin-Dong Kim, Sampo Pyysalo, Sophia Ananiadou, and Pierre Zweigenbaum. 2015. BioNLP Shared Task 2013: Part 1. *BMC Bioinformatics*, 16(Suppl 10), July.

Aurélie Névéol, Kevin Bretonnel Cohen, Cyril Grouin, Thierry Hamon, Thomas Lavergne, Liadh Kelly, Lorraine Goeuriot, Grégoire Rey, Aude Robert, Xavier Tannier, and Pierre Zweigenbaum. 2016. Clinical information extraction at the CLEF eHealth evaluation lab 2016. In *CLEF eHealth Evaluation Lab*.

John P. Pestian, Chris Brew, Pawel Matykiewicz, DJ Hovermale, Neil Johnson, K. Bretonnel Cohen, and Wlodzislaw Duch. 2007. A shared task involving multi-label classification of clinical free text. In *Biological, translational, and clinical language processing*, pages 97–104, Prague, Czech Republic, June. Association for Computational Linguistics.

Hanna Suominen, Sanna Salanterä, Wendy W. Sumitra Velupillai Chapman, Guergana Savova, Noemie Elhadad, Sameer Pradhan, Brett R. South, Danielle L. Mowery, Gareth J.F. Jones, Johannes Leveling, Liadh Kelly, Lorraine Goeuriot, David Martinez, and Guido Zuccon. 2013. Overview of the ShARe/CLEF eHealth evaluation lab 2013. In *Proceedings of CLEF 2013*, Lecture Notes in Computer Science, Berlin Heidelberg. Springer.

Özlem Uzuner, Yuan Luo, and Peter Szolovits. 2007. Evaluating the state-of-the-art in automatic de-identification. *J Am Med Inform Assoc*, 14:550–563.

Özlem Uzuner, Brett R. South, Shuying Shen, and Scott L DuVall. 2011. 2010 i2b2/VA challenge on concepts, assertions, and relations in clinical text. *J Am Med Inform Assoc*, 18(5):552–556, Sep-Oct. Epub 2011 Jun 16.

E Van Mulligen, Z Afzal, S A Akhondi, D Vo, and J A Kors. 2016. Erasmus MC at CLEF eHealth 2016: Concept recognition and coding in French texts. In *CLEF 2016 Online Working Notes*. CEUR-WS.

F. Wingert, David Rothwell, and Roger A Côté. 1989. Automated indexing into SNOMED and ICD. In Jean Raoul Scherrer, Roger A. Côté, and Salah H. Mandil, editors, *Computerised Natural Medical Language Processing for Knowledge Engineering*, pages 201–239. North-Holland, Amsterdam.

Xiaohua Zhou, Xiaodan Zhang, and Xiaohua Hu. 2006. MaxMatcher: Biological concept extraction using approximate dictionary lookup. In *Proceedings of the 9th Pacific Rim International Conference on Artificial Intelligence*, PRICAI'06, pages 1145–1149, Berlin, Heidelberg. Springer-Verlag.

Exploring Query Expansion for Entity Searches in PubMed

Chung-Chi Huang   **Zhiyong Lu**

National Center for Biotechnology Information (NCBI),
National Library of Medicine,
National Institutes of Health (NIH)

chuang@frostburg.edu   Zhiyong.lu@nih.gov

Abstract

Identifying relevant studies from the entire scientific literature is an important task in biomedical research. Past efforts have incorporated semantically recognized biological entities and medical ontologies into biomedical literature search. However, semantic relations are largely overlooked by biomedical search engines. In this work, we aim to discover synonymous biomedical semantic relations between entities and explore their uses in query (semantics) understanding for improved retrieval performance. Specifically, we discover synonymous semantic relations from PubMed queries and apply them to query expansion and specification. In these two real-world scenarios, better PubMed retrieval effectiveness, in terms of recall and precision, can be achieved, demonstrating the utility of our proposed approach.

1 Introduction

PubMed is widely used by millions of users on a daily basis for seeking scholarly publications in biology and life sciences. Recent studies show that a significant portion of PubMed queries are entity specific (i.e. entity searches) (Neveol et al., 2011; Huang and Lu, 2016).

Domain-specific search engines, such as PubMed, typically handle queries with domain knowledge in mind. For example, PubMed incorporates Medical Subject Headings (MeSH) to retrieve documents associated with query's semantic meaning than just keyword matches as in biomedicine it is common for concepts to appear in different forms in user queries and scholarly publications (Lu et al. 2009). However, PubMed can still suffer from mismatches between document and query words when an information need involves entity semantic relations (Baumgartner et al., 2007).

Consider the query *chlorthalidone vs hydrochlorothiazide* and *chlorthalidone versus hydrochlorothiazide*. Semantically similar as they are, PubMed returns twice more relevant documents for the latter, clearly overlooking the semantics of the general terms of *vs* and *versus* during its search. Unfortunately, such performance difference resulting from different query formulations can lead to different levels of user satisfaction and different user experience with PubMed.

In light of this, we propose a framework where we first understand user query's semantics by discovering synonymous patterns among user queries (e.g. patterns *CHEMICAL vs CHEMICAL* and *CHEMICAL versus CHEMICAL*) for entity relations of interest. We then apply these learned synonymous patterns in query expansion to improve retrieval effectiveness for entity searches in PubMed.

In this work, we mine synonymous patterns in user queries instead of scholarly publications because queries are generally short (Islamaj Dogan et al., 2009; Wilkinson et al., 1995) and tend to bond entities in proximity. Here we specifically target chemical-chemical and chemical-disease relations such as chemical-induced-disease relation (Wei et al., 2016). The proposed framework, however, is easily generalizable to understand other bio-entity relations such as protein-protein interaction (Phizicky and Fields, 1995).

Our work is unique in several aspects. First, PubMed queries are semantically analyzed through context patterns, and synonymous relations or synonymous context patterns are discovered automatically. Second, synonymous patterns are applied to expand entity searches at pattern level to improve recall of relevant documents. Third, synonymous patterns can also be applied

Proceedings of the Seventh International Workshop on Health Text Mining and Information Analysis (LOUHI), pages 106–112,
Austin, TX, November 5, 2016. ©2016 Association for Computational Linguistics

to searches with entities only, where we add additional constraints to improve precision. Overall evaluation is able to point key directions for future development and improvement of PubMed, and can also shed light on how to effectively search biomedical literature beyond PubMed.

2   Related Work

Query Expansion (QE) has been an area of active research in Information Retrieval (IR). QE techniques manage to alleviate vocabulary mismatch between query and document words by adding related words to the initial queries, with the goal of improving retrieval effectiveness. Below we discuss three types of QE techniques classified based on how they derive related words: ontology-oriented, query-independent data-driven, and query-dependent data-driven technique.

Ontology-oriented techniques leverage language properties (e.g. synonyms, hypernyms and etc.) in dictionaries (Liddy and Myaeng, 1993), thesauri, or lexical databases (Voorhees, 1994) to find QE. General-purpose lexical database e.g. WordNet (Fellbaum, 1998) or a domain-specific one e.g. MeSH (Nelson et al., 2001) may be used.

Query-independent data-driven QE methods identify queries' similar words by analyzing global-wide documents not specific to queries. Hence, they are also known as global corpus-specific QE methods (Carpineto and Romano, 2012). They learn word association by concept terms (Qiu and Frei, 1993), term clustering (Crouch and Yang, 1992), distributional similarity (Lin 1998; Turney 2001; Chen et al., 2006), semantic topics (Park and Pamamohanarao, 2007), to name a few.

Query-dependent data-driven techniques, on the other hand, analyze query-specific documents for QE. While relevance feedback uses relevant documents from the initial queries, pseudo-relevance feedback uses top-ranked documents without human intervention (Xu and Croft, 1996). Measures for finding related terms in initially returned documents include Rocchio's weighting (Rocchio, 1971), Chi-square (Doszkocs, 1978), and Kullback-Leibler distance (Carpineto et al., 2001). Recently, Cui et al. (2003) and Riezler et al. (2007) consider user-clicked documents relevant for QE.

In biomedicine, QE studies primarily focus on ontologies and pseudo-relevance feedback. For example, Jalali and Borujerdi (2008) and Lu et al. (2009) expand queries via MeSH ontology, and Srinivasan (1996), Aronson (1996), and Zhu et al. (2006) expand queries via Unified Medical Language System (Lindberg et al., 1993). On the other hand, biomedical queries can be reformulated (Lu et al., 2009) or systematically expanded based on initially retrieved documents focusing on abbreviations (Bacchin and Melucci, 2005), the controlled vocabulary of MeSH (Thesprasith and Jaruskulchai, 2014), or open vocabulary (Rivas et al., 2014).

In contrast to previous work, we semantically analyze frequently-sought general patterns (or relations) in biomedical queries, discover pattern synonyms, and use these automatically-learnt synonymous patterns to expand real-world entity searches in PubMed. Such general-phrase pattern-level semantics understanding, complementary to domain-specific MeSH, later proves useful in QE and beneficial to PubMed literature search in our case studies.

3   Entity Searches in PubMed

(a) PubMed titles for the search *midazolam sevoflurane*
1. Network Meta-Analysis on the Efficacy of Dexmedetomidine, **Midazolam**, Ketamine, Propofol, and Fentanyl for the Prevention of **Sevoflurane**-Related Emergence Agitation in Children.   2. Determination of optimum time for intravenous cannulation after induction with **sevoflurane** and nitrous oxide in children premedicated with **midazolam**
(b) PubMed titles for its semantics-constrained query *midazolam vs sevoflurane* OR *midazolam versus sevoflurane* OR …
1. Long-term sedation in intensive care unit: a randomized **comparison** between inhaled **sevoflurane** and intravenous propofol or **midazolam**.   2. Complications of **sevoflurane**-fentanyl **versus midazolam**-fentanyl anesthesia in pediatric cleft lip and palate surgery: a randomized comparison study.

Table 1. An example of PubMed search results (sorted by relevance) without (a) and with (b) semantic expansion.

We focus on understanding users' information needs or search semantics when they submit entity searches to PubMed. We discover synonymous patterns or entity relations in user queries (Section 3.1) and exploit them in the following two use scenarios to improve PubMed retrieval effectiveness.

Scenario 1. Consider an entity pair search with explicit relation mention (e.g. comparison relation between two drugs as in *albuterol vs levalbuterol*). We expand the query with its synonymous counterparts belonging to the same pattern-level relation (e.g. adding *albuterol versus levalbuterol*, *comparison between albuterol and levalbuterol*, and etc.). With such query expansion, we expect to retrieve

semantic relation	pattern	BiR	semantic relation	pattern	BiR
drug comparison	*#C versus #C*	2.38	drug-induced-disease	*#C induced #D*	1.14
	#C vs #C	10.05		*#D induced by #C*	1.14
	comparison of #C and #C	1.91		*#D associate with #C*	969.6
	comparison #C and #C	1.91		*#C side effect #D*	303
	#C compare #C	135.65		*#D caused by #C*	21.07
	difference between #C and #C	144		*#C exposure and #D*	21.26
	comparison between #C and #C	1.91		*#C cause #D*	48
	#C compare to #C	135.65		*#D risk factor #C*	94.13
	#C compare with #C	135.65		*#D #C adverse effect*	484.8
	difference #C and #C	144			
drug combination	*#C and #C combination*	1.35	drug-treats-disease	*#D treatment #C*	1.96
	combination of #C and #C	1.35		*#D and #C therapy*	2.41
	combine #C and #C	904.2		*treatment of #D with #C*	1.96
	#C in combination with #C	1.35		*treatment of #D #C*	1.96
	#C and #C combination therapy	6.14		*#D treatment with #C*	1.96
	#C combined with #C	4.93		*#C treatment for #D*	1.96
	add #C to #C	37.99		*#C in the treatment of #D*	1.96
	combination therapy with #C and #C	6.14		*#C in #D treatment*	1.96
	concomitant #C and #C	38.64		*#D treated with #C*	7.59
				#C therapy in #D	2.41

Table 2. Retrieval benefit in recall (**BiR**) when using synonymous relational patterns for query expansion.

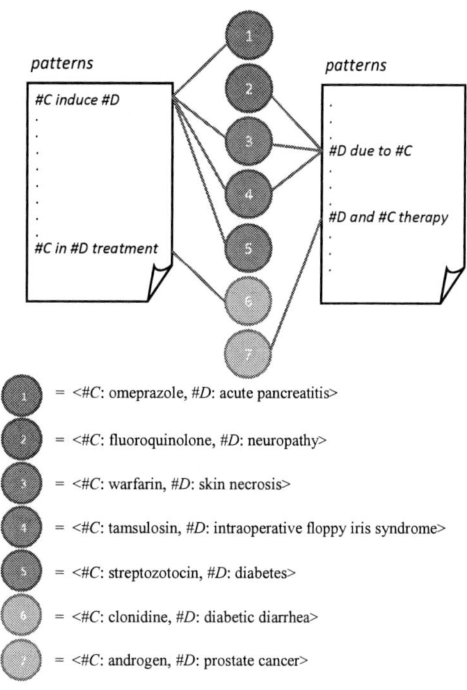

Figure 1. Patterns' semantics similarity in terms of overlapping entities or LSA topics. While circles represent entities, the colors of the circles represent learned LSA topics.

from PubMed additional documents originally unreachable and expect to balance PubMed results across different query formulations with identical semantics meaning.

Scenario 2. Consider a pure entity pair search without any explicit mention of entity relation (e.g. *midazolam sevoflurane*). We constrain the query on its known search semantics learned based on past PubMed searches (e.g. comparison relation between these two drugs). The newly constructed search (e.g. *midazolam vs sevoflurane* OR *midazolam versus sevoflurane* OR ... where OR combines Pub-Med results) is expected to direct PubMed towards documents users truly interested in but otherwise might be ranked low based on the original search. Take Table 1 for example. Top-ranked documents are more relevant with the new semantics-constrained query if users are to compare the two entities without explicitly mentioning so in the search query.

Note that in *Scenario 1* and *Scenario 2* we perform PubMed searches under relevance sorting (as opposed to the default chronical sorting) and we search PubMed and use matches in article titles as a proxy for human relevance evaluation (Kim et al., 2016). In other words, to ensure quick turnaround and large-scale evaluation, we assume those matching titles all satisfy users' information needs (i.e. perfect precision) and thus no human relevance judgments is required.

3.1 Discovering Synonymous Patterns

We have previously developed an unsupervised approach for identifying synonymous patterns of entity relations in PubMed queries (Huang and Lu, 2016). Due to space limitation, we only briefly outline major steps below. We refer interested readers to (Huang and Lu, 2016) for details.

First, a six-month worth of PubMed queries (35M queries) are stemmed and tagged using entity recognition tools (Wei et al., 2015;

Leaman et al., 2013; Leaman et al., 2015) for genes/proteins, diseases, and chemicals/drugs.

Next, we formulate queries to context patterns and focus on specifically discovering synonymous patterns for chemical-chemical (*CC*) and chemical-disease (*CD*) relations. For instance, the query *skin necrosis associate with warfarin* is formulated into *#D associate with #C* where *#C* and *#D* stands for chemical and disease entity respectively.

Inspired by distributional similarity (Lin 1998), we then exploit these patterns' participating entity pairs to understand their semantics. In such a way, synonymous patterns can be found in an unsupervised fashion in contrast to seeds-required pattern recognition work (e.g. Xu and Wang, 2014). Take Figure 1 for example. Our framework will consider the pattern *#C induce #D* semantically closer to *#D due to #C* than to *#C in #D treatment* since *#C induce #D* and *#D due to #C* share more participating entities in user queries: 2 overlapping entities out of 7 entities vs 0 out of 7.

To avoid data sparseness issue on (distributional similarity in) entity mention, we further leverage latent semantic analysis, LSA, (Rehurek and Sojka, 2010) to find entities' LSA topics which in turn reduces the space of semantics analysis from the dimension of entity pairs to a much smaller dimension of LSA topics. The benefit of using LSA topics is clear: after LSA transformation, *#C induce #D* in Figure 1, where circle's colors depict LSA topics, shows stronger semantics connection with *#D due to #C* than previously without LSA: 2 overlapping LSA topics out of 3 topics.

Our LSA-based approach is able to achieve satisfying performance in finding semantically similar patterns across entity relations of interest, such as drug-induced-disease relation, drug-drug interaction, to name a few. We refer interested readers to (Huang and Lu, 2016) for detailed evaluation results.

3.2 Expanding Entity-Relation Searches

Once our method identifies candidates of pattern synonyms, we collect the set of true synonymous patterns and apply them to semantic query expansion as below.

We first order a semantic relation's synonymous patterns according to their frequencies in PubMed queries, which represent user preferences or user intuitions (in searching the target bio-relation between two entities). See patterns in descending order of frequency in the second and fifth column of Table 2. For example, Pub-Med users prefer using *#C versus #C* to *#C vs #C* or *comparison of #C and #C* in comparing two drugs. Currently, four common entity relations between drugs and between drugs and diseases are of our particular interest: drug comparison, drug combination, drug-induced-disease and drug-treats-disease.

Second, for each relation, we assemble its 500 most searched entity pairs from our search logs. For example, <albuterol, levalbuterol> is a popular chemical pair for the drug comparison relation.

For each entity pair (e.g. <albuterol, levalbuterol>) of a semantic relation, we then submit a query with the pair using one of the relational patterns (e.g. *albuterol vs levalbuterol*) and compare the search result with that of semantically expanded query that leverages all synonymous patterns (e.g. *albuterol versus levalbuterol* OR *albuterol vs levalbuterol* OR ... Syntax OR combines PubMed retrieval results). Recall that the searches are limited to PubMed titles. Finally, we compute the ratio of the number of total search results via all patterns of the semantic relation over that of each individual pattern, averaged over 500 entity pairs. Such difference in recall is referred to as benefit in recall, **BiR**.

As Table 2 shows, a **BiR** score above 1 means expanding queries using collective synonymous patterns of the same semantics improves PubMed recall or helps PubMed retrieve more relevant documents. Take the drug comparison relation for example. Regardless of the chemical pair of interest, expanded queries can always retrieve more relevant documents than using the individual pattern of *#C versus #C* (more than twice as many on average: 2.38). In some cases of Table 2, the improvement in recall is substantial (e.g. 135.65 associated with *#C compare #C*, 904.2 associated with *combine #C and #C*, and so on).

The benefit of using our synonymous patterns for query expansion in current PubMed settings can be observed across various types of *CC* or *CD* entity-relation searches, searches with explicit relation mention. And interestingly, the most frequently used patterns by users (or the most intuitive/straightforward search patterns from users' points of view) may not always be the best choice at default: among the drug comparison patterns, *comparison of #C and #C* is more effective than the most popular *#C versus #C* in retrieving relevant documents. A semantic framework like ours can balance PubMed retrieval results across different entity-relation expressions in searches with similar meanings.

3.3 Expanding Pure Entity Pair Searches

Among PubMed searches, pure entity pair searches or searches containing only two bio-entities without any explicit relation mentions (e.g. *midazolam sevoflurane*), account for approximately half of the searches involving dual bio-entities. As a result, we investigate in this subsection how we can improve PubMed user experience by expanding these queries, with the help of our synonymous patterns and past user searches. The process is detailed below.

First, we identify pure entity pair searches only sought by PubMed users in *a specific* relation/context, based on which we expand the searches and impose semantic search constraints. Take the pure entity pair search *midazolam sevoflurane* for instance. Since it had only been searched with drug comparison relation by PubMed users, we later explicitly constrain that search query in the context of drug comparison relation. This step infers the implicit relation between the entity pair from the wisdom of the crowd (i.e. past search logs). Our hypothesis is that such implicit relation, if explicitly added to the search, may improve retrieval results and in turn user experience.

In the current experiment, a total of 1,600 unique pure entity-pair queries are collected with *CC* relation constraints (i.e. drug comparison, drug combination, and drug interaction) and *CD* relation constrains (i.e. drug-treats-disease, drug-induced-disease, supplement-for-disease, drug-resistance-in-disease).

Similar to the settings in Section 3.2, we submit to PubMed (a) original queries, i.e. pure entity pairs and (b) expanded queries with explicit relation constraints learnt from past user queries. For example, original search *midazolam sevoflurane* and its semantics-constrained counterpart *midazolam versus sevoflurane* OR *midazolam vs sevoflurane* OR ... (expanded using our synonymous patterns of the drug comparison relation, in which *midazolam sevoflurane* had only been sought) will be submitted to PubMed.

Finally, based on the search results from (a) and (b), we compute the retrieval effectiveness of regular PubMed by using (b)'s results as the ground truth. In other words, we assume the expanded queries truly represent users' search intention and their search results truly satisfy users' information needs. Retrieval performance is measured by standard information retrieval (IR) measures: precision (P), mean reciprocal rank

(MRR) and nDCG (Jarvelin and Kekalainen, 2002) at rank 20.

As we can see in Table 3, the difference between current performance scores in MRR or nDCG and perfect scores (i.e. perfect MRR or nDCG equals 1) suggests genuinely there is room for performance increase in retrieval for such searches, i.e. pure entity pair searches, in current PubMed settings. While pure *CD* searches yield better results than pure *CC* searches, potential gain in performance is still substantial for *CD* queries, which can be achieved by simply adding semantics constraints and expanding queries accordingly. In some cases (e.g. pure entity pair searches with implicit drug interaction relation), semantics constraints almost warrant a more satisfying search performance.

entity pair type	implicit relation	IR measures @ 20	results
CC	drug comparison	*P*	0.25
		MRR	0.43
		nDCG	0.57
	drug combination	*P*	0.29
		MRR	0.47
		nDCG	0.61
	drug interaction	*P*	0.13
		MRR	0.32
		nDCG	0.43
CD	drug-treats-disease	*P*	0.34
		MRR	0.58
		nDCG	0.66
	drug-induced-disease	*P*	0.36
		MRR	0.63
		nDCG	0.70
	supplement-for-disease	*P*	0.23
		MRR	0.47
		nDCG	0.56
	drug-resistance-in-disease	*P*	0.21
		MRR	0.43
		nDCG	0.55

Table 3. Results on pure *CC* and *CD* queries with implicit relations.

4 Summary

We have applied query semantics understanding to PubMed literature search. The proposed framework involves discovering synonymous relational patterns in queries and, based on those, expanding PubMed user queries, specifically entity search queries. Preliminary evaluation shows such semantic query expansion helps to improve PubMed retrieval effectiveness. And better PubMed performance implies better user experience and less curation effort (Lu and Hirschman, 2012). Incorporating such general-phrase semantics framework, complementary to domain-specific MeSH, into PubMed serving millions of users is warranted.

5   Acknowledgements

This work was supported by the Intramural Research Program of the National Library of Medicine, National Institutes of Health. The authors would like to thank anonymous reviewers for their suggestions and comments.

Reference

Aronson, A.R. 1996. The effect of textual variation on concept based information retrieval. *Proc AMIA Annu Fall Symp.*

Aronson, A.R. and T.C. Rindflesch. 1997. Query expansion using the UMLS Metathesaurus. *Proc AMIA Annu Fall Symp.*

Bacchin, M. and M. Melucci. 2005. Symbol-based query expansion experiments at TREC 2005 Genomics Track. In *Proceedings of Text REtrieval Conference.*

Baumgartner, W, Z. Lu, H. Johnson et al. 2007. An integrated approach to concept recognition in biomedical text. In *Proceedings of BioCreative Challenge Evaluation Workshop.*

Carpineto, C., R. De Mori, G. Romano et al. 2001. An information-theoretic approach to automatic query expansion. *ACM Transactions on Information Systems.*

Carpineto, C. and G. Romano. 2012. A survey of automatic query expansion in information retrieval. *ACM Computing Surveys.*

Chen, H., M. Lin, and Y. Wei. 2006. Novel association measures using web search with double checking. In *Proceedings of ACL*, p. 1009-1016.

Crouch, C.J. and B. Yang. 1992. Experiments in automatic statistical thesaurus construction. In *Proceedings of ACM SIGIR.*

Cui, H., J.R. Wen, J.Y. Nie et al. 2003. Query expansion by mining user logs. *IEEE Transactions on Knowledge and Data Engineering.*

Deerwester, S., S.T. Dumais, G.W. Furnas et al. 1990. Indexing by latent semantic analysis. *Journal of the Association for Information Science.*

Diaz-Galiano, M.C., M.T. Martin-Valdivia, and L.A. Urena-Lopez. 2009. Query expansion with a medical ontology to improve a multimodal information retrieval system. *Comput Biol Med.*

Dramé, K., F. Mougin, and G. Diallo. 2014. Query expansion using external resources for improving information retrieval in the biomedical domain. In *Proceedings of ShARe/CLEF eHealth Evaluation Lab.*

Doszkocs, T.E. 1978. AID, an associative interactive dictionary for online searching. *Online Review.*

Fellbaum, C. 1998. WordNet: an electronic lexical database.

Gauch, S., J. Wang, and S.M. Rachakonda. 1999. A corpus analysis approach for automatic query expansion and its extension to multiple Databases. *ACM Transactions on Information Systems.*

Gonzalo, J., F. Verdejo, I. Chugur et al. 1998. Indexing with WordNet synsets can improve text retrieval. In *Proceedings of ACL Workshop.*

Huang, C.C. and Z. Lu. 2016. Discovering biomedical semantic relations in PubMed queries for information retrieval and database curation. *Database.*

Islamaj Dogan, R., G.C. Murray, A. Neveol et al. 2009. Understanding PubMed user search behavior through log analysis. *Database.*

Jalali, V. and M.R.M. Borujerdi. 2008. The effect of using domain specific ontologies in query expansion in medical field. In *Proceedings of IEEE International Conference on Innovations in Information Technology.*

Jarvelin, K. and J. Kekalainen. 2002. Cumulated gain-based evaluation of IR technologies. *ACM Transactions on Information Systems.*

Kim, S., W.J. Wilbur, Z. Lu. 2016. Bridging the gap: a semantic similarity measure between queries and documents. arXiv:1608.01972.

Lavrenko, V. and W.B. Croft. 2001 Relevance based language models. In *Proceedings of ACM SIGIR.*

Leaman, R., R. Islamaj Dogan, and Z. Lu. 2013. DNorm: disease name normalization with pairwise learning to rank. *Bioinformatics.*

Leaman, R., C.H. Wei, and Z. Lu. 2015. tmChem: a high performance approach for chemical named entity recognition and normalization. *J Cheminform.*

Liddy, E.D. and S.H. Myaeng. 1993. DR-LINK's linguistic-conceptual approach to document detection. In *Proceedings of Text REtrieval Conference.*

Lin, D. 1998. Automatic retrieval and clustering of similar words. In *Proceedings of ACL*, p. 768-774.

Lindberg, D.A., B.L. Humphreys, and A.T. McCray. 1993. The Unified Medical Language System. *Methods Inf Med.*

Lu, Z. and L. Hirschman. 2012. Biocuration workflows and text mining: overview of the BioCreative 2012 Workshop Track II. *Database.*

Lu, Z., W. Kim, and W.J. Wilbur. 2009. Evaluation of Query Expansion Using MeSH in PubMed. *Inf Retr.*

Lu, Z., W.J. Wilbur, J.R. McEntyre et al. 2009. Finding query suggestions for PubMed. In *AMIA Annu Symp Proc.*

Nelson, S.J., W.D. Johnston, and B.L. Humphreys. 2001. Relationships in medical subject headings (MeSH).

Neveol, A., R. Islamaj Dogan, and Z. Lu. 2011. Semi-automatic semantic annotation of PubMed queries: a study on quality, efficiency, satisfaction. *J Biomed Inform.*

Park, L.A.F. and K. Ramamohanarao. 2007. Query expansion using a collection dependent probabilistic latent semantic thesaurus. In *Proceedings of PAKDD.*

Phizicky, E.M. and S. Fields. 1995. Protein-protein interactions: methods for detection and analysis. *Microbiol Rev.*

Qiu, Y. and H.P. Frei. 1993. Concept based query expansion. In *Proceedings of ACM SIGIR.*

Rehurek, R. and P. Sojka. 2010. Software framework for topic modelling with large corpora. In *Proceedings of LREC Workshop.*

Riezler, S., E. Vasserman, I. Tsochantaridis et al. 2007. Statistical machine translation for query expansion in answer retrieval. In *Proceedings of ACL.*

Rocchio, J.J. 1971. Relevance feedback in information retrieval.

Srinivasan, P. 1996. Query expansion and MEDLINE. *Information Processing and Management.*

Thesprasith, O. and C. Jaruskulchai. 2014. Query expansion using medical subject headings terms in the biomedical documents. *Intelligent Information and Database Systems.*

Tsuruoka, Y. and J. Tsujii. 2005. Bidirectional inference with the easiest-first strategy for tagging sequence data. In *Proceedings of EMNLP*, p. 467-474.

Turney, P.D. 2001. Mining the Web for synonyms: PMI-IR versus LSA on TOEFL. In *Proceedings of EMCL*, p. 491-502.

Voorhees, E.M. 1994. Query expansion using lexical-semantic relations. In *Proceedings of ACM SIGIR.*

Wei, C.H., H.Y. Kao, and Z. Lu. 2015. GNormPlus: an integrative approach for tagging genes, gene families, and protein domains. *Biomed Res Int.*

Wei, C.H., Y. Peng, R. Leaman et al. 2016. Assessing the state of the art in biomedical relation extraction: overview of the BioCreative V chemical-disease relation (CDR) task. *Database.*

Wilkinson, R., J. Zobel, and R. Sacks-Davis. 1995. Similarity measures for short queries. In *Proceedings of Text REtrieval Conference.*

Xu, R. and Q. Wang. 2014. Automatic construction of a large-scale and accurate drug-side-effect association knowledge base from biomedical literature. *J Biomed Inform.*

Xu, X. and X. Hu. 2010. Cluster-based query expansion using language modeling in the biomedical domain. In *Proceedings of IEEE International Conference on Bioinformatics and Biomedicine Workshops.*

Zhai, C. and J. Lafferty. 2001. Model-based feedback in the language modeling approach to information retrieval. In *Proceedings of CIKM.*

Zhu, W., X. Xu, X. Hu et al. 2006. Using UMLS-based re-weighting terms as a query expansion strategy. In *Proceedings of IEEE International Conference on Granular Computing.*

Author Index

Bakewell, Robert, 17
Bernstam, Elmer, 43
Bethard, Steven, 37

Chalapathy, Raghavendra, 1
Cohen, Kevin, 78
Cohen, Trevor, 43
Collier, Nigel, 85
Cornegruta, Savelie, 17

D'hondt, Eva, 61

Erk, Katrin, 86

Ferracane, Elisa, 86

Grau, Brigitte, 61
Grouin, Cyril, 61, 78

Huang, Chung-Chi, 106

Johnson, Todd, 43

Kirchhoff, Katrin, 52
Kokkinakis, Dimitrios, 28

Lalor, John, 69
Lavergne, Thomas, 96
Lu, Zhiyong, 106
Lundholm Fors, Kristina, 28

Marshall, Iain, 86
Montana, Giovanni, 17
Mukherjee, Arjun, 37
Munkhdalai, Tsendsuren, 69

Neveol, Aurelie, 78
Nordlund, Arto, 28

Pedersen, Ted, 37
Petersen, Steffen, 6

Piccardi, Massimo, 1

Rey-Villamizar, Nicolas, 37
Riedel, Sebastian, 6
Robert, Aude, 78

Sadeque, Farig, 37
Shrestha, Prasha, 37
Solorio, Thamar, 37
Spithourakis, Georgios, 6

Turner, Anne M., 52

Wallace, Byron, 43
Wallace, Byron C., 86
Withey, Samuel, 17

Yu, Hong, 69
Yu, Zhiguo, 43

Zare Borzeshi, Ehsan, 1
Zweigenbaum, Pierre, 96

Association for Computational Linguistics
209 N. Eighth Street
Stroudsburg, Pennsylvania 18360

ISBN 978-1-5108-3152-0